Desert Problems and Desertification in Central Asia

Springer

Berlin
Heidelberg
New York
Barcelona
Hong Kong
London
Milan
Paris
Singapore
Tokyo

Agajan G. Babaev (Ed.)

Desert Problems and Desertification in Central Asia

The Researches of the Desert Institute

With 60 Figures, 21 in Color, and 37 Tables

Springer

EDITOR

Professor Dr. Agajan G. Babaev
Desert Institute
15 Bitarap Turkmenistan Street
744 000 Ashgabat
Turkmenistan
e-mail: babaev@desert.ashgabat.su

SCIENTIFIC ADVISER

Professor Dr. Josef E. Breburda
GEOPLAN

LAYOUT AND TECHNICAL EDITING

Joachim Langbein
GEOPLAN

PROJECT IMPLEMENTATION

Dr. Ernst Klimm
GEOPLAN
Forststraße 7
D-50226 Frechen
Germany
e-mail: geoplan.klimm@t-online.de

ENGLISH LANGUAGE EDITING

Tim Jacob Gluckman
Im Fichtenhain 19
D-51069 Köln
Germany

This book has been elaborated within the German-Turkmen co-operation project „Desertification Combat in Turkmenistan".

ISBN 3-540-65647-2 Springer-Verlag Berlin Heidelberg New York

Library of Congress Cataloging-in-Publication Data

Desert Problems and Desertification in Central Asia: the researches of the Desert Institute /Editor: Agajan G. Babaev. - Berlin; Heidelberg; New York; Barcelona; HongKong; London; Milan; Paris; Singapore; Tokyo: Springer, 1999
ISBN 3-540-65647-2

Cover Design: Erich Kirchner, Heidelberg
Typesetting: Camera-ready by Joachim Langbein, GEOPLAN

SPIN: 10713922 30/3136 – 5 4 3 2 1 0 – Printed on acid free paper

Preface

The aim of this publication is to accord the international donor community and a wide expert audience access to the experience gathered in Central Asia under the auspices of the Ashgabat Desert Institute in researching and tackling the problems of arid areas. In the International Convention to Combat Desertification (CCD), the process of learning from the past and systematically collating experience is made into a fundamental principle. The significance of this publication, therefore, is that it collates for the first time the store of experience amassed by the Desert Institute and makes it available to an international audience.

The publication is a resource to be used in international and interdisciplinary co-operation in the field, particularly against the background of the massive ecological damage caused by the Soviet Union in Central Asia. It is thus also intended to help preserve the expertise that has been developed over the last few decades in the Desert Institute so that it can be drawn on in future in efforts to remedy the ecological damage in the region. What makes the calls for co-operation in this area so important is the danger of a fragmented approach emerging as result of the number of different political centers and the variety of assistance provided by outside donors. The problems that have to be tackled are too great to allow isolated approaches to tackling the problem in Central Asia.

I am delighted that, through the Federal Ministry for Economic Co-operation and Development, the Federal Republic of Germany was able to play a large part in the publishing of this work. Germany is, incidentally, now also home to the Secretariat of the UN Convention to Combat Desertification whose headquarters are in Bonn. All of us, the countries of Central Asia and the countries of the world that wish to see them achieve progress, are faced with the task of checking a natural disaster which has been caused by human hubris. This we must do in the interests of humankind as a whole. May this publication play a part in the process. I hope that it will be devoted the attention it deserves by politicians and the academic community, who are charged with tackling the task.

Martin Lehman
Head of the Central Asia and Caucasus Division of the
Federal Ministry for Economic Co-operation and Development

Contents

1. Introduction

A.G. Babaev

At the junction of the 20th and 21st centuries relations between people and nature have become more complicated, reached an enormous scale and turned into a real threat to the intactness of the whole ecological system. Today environmental issues are involved, in various forms and to different degrees, in all most critical aspects of present-day reality.

A complicated and multifaceted plan for the optimization of the natural environment is essential for desert territories covering as they do more than 30% of the land area on the Earth. The natural environment of the desert is highly vulnerable and fragile, and that is why any unwise action by the human species in resources development leads to a disturbance of equilibrium in nature, and accelerates desertification processes.

Desertification is a result of a complex interaction between a socio-economic system and natural factors. These factors may act as avalanches intensifying each other.

The desertification problem was shaping in the period of the very dramatic events of 1968-1973 when severe and long drought affected the Sahelian zone south of the Sahara in Africa. And ever since desertification processes have continued to develop and widen their boundaries from year to year. Now it can be stated with certainty that the problem "people-desert-desertification" has reached the status of a "global issue" having serious, desertification-related consequences in the ecological, social, economic and political spheres.

It is known that desert territories possess rich natural and labor resources, so their rational, stage-by-stage development should be considered a major, strategic reserve for the development of the society. But instead in arid areas mass suffering due to starvation and diseases can be observed, along with clashes of the most critical kind in economic, social and ecological matters. In these conditions any scientific, practical and pecuniary aid to people who find themselves in an intensive, desertification area is the humane duty of the world community.

The problem of desertification control at a global level was first discussed at the UN Conference on Desertification (UNCOD) in 1977 in Nairobi. This Conference gave a retrospective analysis and outlined steps to implement the "Plan of Action to Combat Desertification" adopted at it. From the very beginning, this problem was regarded by the scientific community as one of the most serious issues involving the interests of countries and peoples not only in the desert zone, but in adjoining territories, too. Within the framework of the search for practical solutions, a taskforce was formed to transfer available experience in research and development of resources of desert lands to the countries affected by desertification. That conference approved a proposal to set up the Regional Center for Research and Personnel Training in Desertification on the basis of the Desert Institute of Turkmenistan (Ashgabat) within the framework of the United National Environmental Program (UNEP) and the UN Economic and Social Commission for Asia and the Pacific (ESCAP). Implementation of several programs started in 1979, when Permanent Seminars in Central Asia and the Caspian region were organized.

During the time of its existence the Desert Institute has finished a number of integrated researches, and developed and provided hundreds of recommendations of great economic significance. The results of its scientific researches are regularly published as monographs, thematic scientific treatises and in journals. Beginning in 1967 the Institute has edited the international, theoretical, scientific journal "*Problems of Desert Development*" which since 1980 has been published in English by the Allerton Press in the USA, and is widely distributed throughout the world.

Wide use of the natural, economic potential of arid territories for development of countries and peoples is possible only if complex research and the integrated development of desert resources are practiced from and within the framework of general socio-economic progress. This was stressed at the UN International Conference on Environment and Development – the "Earth summit" held in 1992 in Rio de Janeiro. It is important to borrow everything rational from the old, traditional methods of development, modern technology ensuring that unfavorable consequences for the natural environment are minimized. The leading strategic principle of desertification control is not preservation of intact arid regions, but rather their development and enrichment with a view to improving living standards of the local populations.

Today the Desert Institute of Turkmenistan actively cooperates with other countries in desertification combat, and so the Institute has managed to widen enormously understanding of the mechanisms of degradation of the desert environment, to carry out an in-depth analysis of world approaches to desertification control, to exchange experiences on methods of study of regional environmental specifics of desertification processes, and to transfer integrated experience on development of desert territories to countries of the global arid zone.

A significant place in the consulting activity of the Institute is given to the implementation of research work within the framework of International Projects. These projects widely use results of fundamental researches, scientific methodological developments, and experiences of the Desert Institute over many years.

This book describes the results of researches and tests conducted by the Desert Institute of Turkmenistan in arid lands of Central Asia. Throughout all this work, the contributors to the book profoundly studied and used the wise experience of local populations in careful use of nature and their activities in desert conditions.

The authors are leading scientists of the Desert Institute of Turkmenistan. They describe most explicitly the results of both their theoretical and practical works on desert problems and desertification control.

The publication of the book is financed by the German Agency for Technical Cooperation (GTZ). The aim of the GTZ support is to concentrate the practical scientific experience of the Desert Institute towards the solution of this important problem. It is hoped that many national and regional centers will take interest in the labors of the Desert Institute all of which are connected with the realization of the UN Convention to Combat Desertification.

2. The Natural Conditions of Central Asian Deserts

A.G. Babaev

ABSTRACT: A wide ranging overview is given of the natural environment in Central Asian deserts. The geology and geomorphology includes such features as deflation and barkhans whose genesis has been of interest throughout the century. The precipitation and winds are specified along with other climatic features. Surface water is of major relevance to studies of desertification; an unusual feature here are the sophisticated, traditional modes of storage in reservoirs and underground canals. Soil composition is usually either eolian or salinated, and hence unproductive. The very varied flora is unusually dependent on the environment; the predominant plant communities listed.

Deserts of Central Asia, mostly covering lowlands, extend from the eastern coast of the Caspian Sea to the piedmonts of the Tien-Shan and Altai. The area of the deserts consists of ca. 250 million hectares.

Notwithstanding similarities of climatic conditions and regularities of formation, the desert landscapes vary strongly above all in their geological structure, soil and vegetation cover according to their geographical location.

2.1 Geology and Geomorphology

The northern regions of Central Asia are the most ancient. They are composed of Paleozoic deposits with inclusions of crystalline rocks. Ancient mountain structures have been gradually destroyed, and leveled maintaining in some parts the slightly broken relief of a stone desert.

Geologically much younger are western, central and southern parts of a desert belt where the Tertiary and Quarternary deposits dominate. Layers of sedimentary rocks (limestones, marls, sandstones, clays and others) interbed in various combinations.

Many geological surveys have taken part in studies of the Turanian lowland surface. Deep drilling has revealed a complex of sedimentary rocks formed in the post-Jurassic period in Central Asian deserts. A sequence of their facies proves that there were active tectonic dislocations of the earth's crust not only in the Early Paleozoic, but also in the Mesozoic. This finally led to displacement of the areas of outcrop, transportation and accumulation of material. Study of sedimentary rocks and their accumulation conditions enables the formation of a paleogeographical picture of the studied territory over a long geological period.

Continental Pliocene and Quarternary deposits are widespread over the whole territory of the Central Asian deserts. During the Quarternary period, sediments accumulated as a result of the action of ancient rivers in the conditions of a generally continental, desert regime. In periods of increased water flow in rivers as a result of melting of glaciers in the mountains, wandering streams were formed. The activity of ancient streams is clearly visible in the modern relief of piedmont areas and plains.

Erosion processes are most intensive in piedmont areas composed mainly of poorly cemented sandstones, loess loams and similar ground subject to easy scouring and weathering. After the demise of the ancient river system, the alluvial plains were gradually subject to deflation and eolian dissection.

Therefore, in Central Asia most widespread are sandy deserts that were largely formed in areas of development of ancient or modern alluvial or lacustrine-marine, loose deposits. These are, for the most part, parts of huge alluvial and deltaic plains over which in the Neogene-Quarternary period rivers meandered, and marine transgressions developed. And it is the ancient river systems and marine transgressions that led to the formation of thick sandy deposits. Further on these loose sandy deposits were subject to eolian processes as a result of which different forms of sandy relief were shaped. In the deserts a degree of dissection of the initially flat surface depends on the wind force, and the period of its action on the initial plain surface. (Fig. 2.1.).

Intensity of eolian transformation of top horizons in sandy deserts is dependent, first of all, on a wind regime and soil lithology. Consequently, modern desert relief has been shaped by the conditions of a dry, continental climate over a long period. In the course of denudation, transportation and accumulation of mostly Quarternary sand deposits, the principal processes were alluvial while eolian processes were responsible for a radical transformation of flat topography. In some parts of deserts the products of eolian transformation were Pliocene alluvial deposits, Quarternary alluvial deposits, deposits of marine transgressions, or deltaic deposits. This is proved by a similarity of the mineralogical composition, and coloring of grains of eolian and original underlying sands.

The principal, relief forming factor in a sandy desert, both in the earlier geological periods and today, is wind: it causes deflation, transfer and accumulation of sand. Eolian processes that go on in a sandy desert for a long time have broken the initially loose sands and formed a present-day eolian relief that has acquired high stability in the wind action.

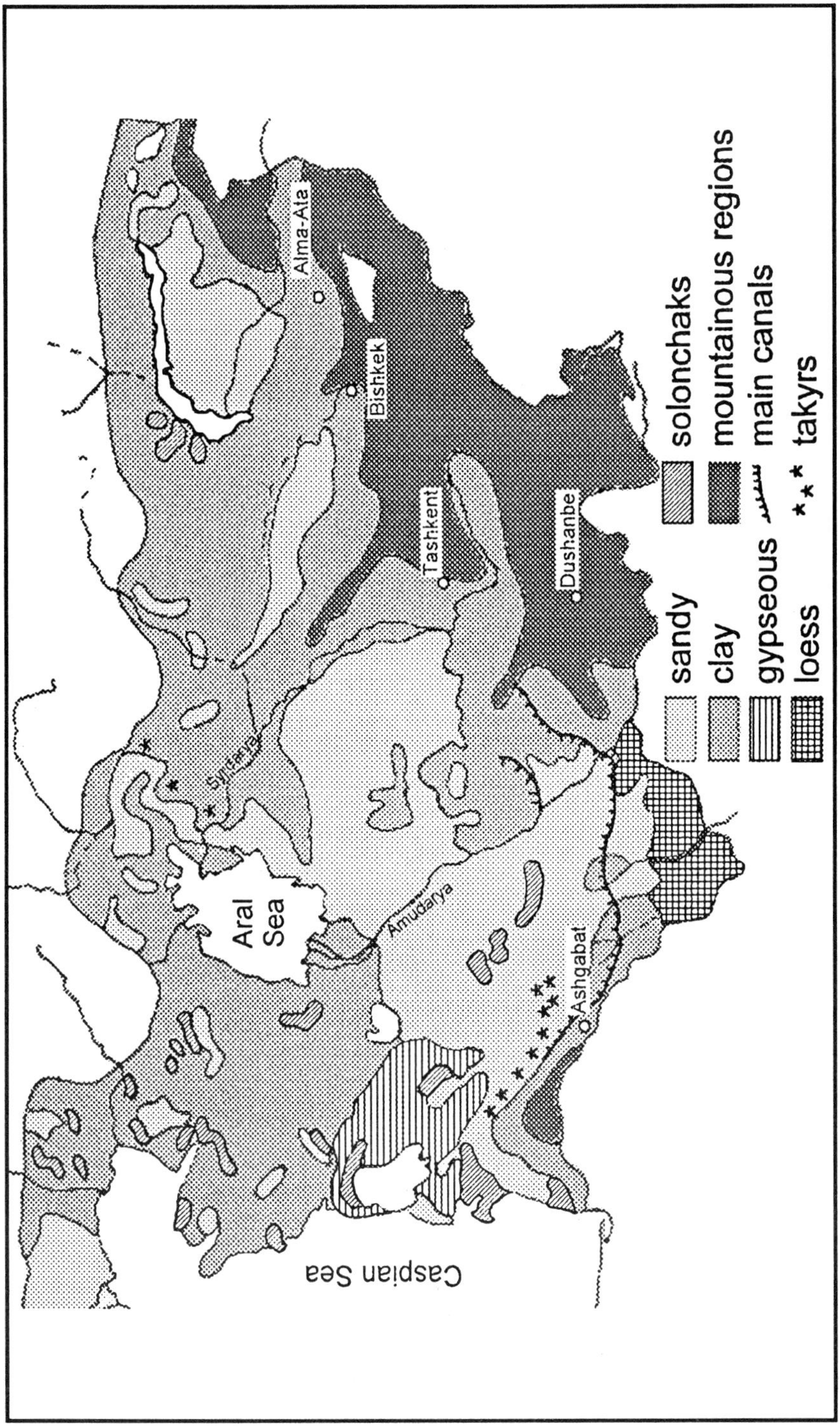

Fig. 2.1. Desert Types of Central Asia

However, relative stability of modern eolian forms hardly means that eolian processes stopped developing. This is proved by vast expanses of broken moving sands where modern relief-forming processes are most intensive. Here constant re-deflation of loose, sand material is observed. It depends on wind force and the physical state of the ground. Eolian material moves with a wind-sand flow (storm of drifting sand), and with mobile eolian forms (barkhans, chains, dunes) when the wind speed reaches 4 m/s and more at a windvane height. Direction and rate of transfer are dependent on a wind regime in a particular desert territory.

On eolian, relief forms covered with vegetation the deflation processes reveal themselves differently and are very slow, thus, creating an impression of complete unchangeability of these forms. Surfaces of overgrown territories under natural conditions (not affected by anthropogenic factors) are subject to a varying degree of deflation. But in general, they are characterized by higher stability due to the presence of vegetation. A slightly broken surface of sand deposits is subject to deflation to a lesser extent. For instance, on smoothly rolling sands there are usually no clear-cut traces of outcrop, transfer and build-up of eolian material. Deflation on severely broken sand surfaces is different on different relief elements. On sandy ridges the lower part of slopes and inter-ridge depressions are subject to slight drifting owing to a dense grass cover, floors caused by a high level of silty particles in sands and the shallow occurrence of poorly compacted primary deposits or groundwater aquifers: all of these could be local bases of deflation. The tops of sand forms and the upper parts of their slopes have traces in the form of ulcers or even barkhans. Deflation on tops of ridges becomes more active due to stronger drying-out of sands and a deeper wind impact.

Therefore, the main forms of sand relief in Central Asian deserts are the ridge parallel to each other. They are very elongated but underdeveloped vertically and in the basement. Lengths of sandy ridges varies from several hundred meters to several kilometers, and their width from 10-20 m to a kilometer, sometimes even more, and a height from 5-6 m to 60 m. Their parameters in different parts of deserts are different. When long and high ridges are joined together in some places by perpendicular, short and low ridges, they all form a honeycomb-like relief.

Sand in ridges is stabilized by plants, and that is why they do not move. The upper parts of ridges often resemble barkhans. In sandy deserts hillocky sands from 2 to 20-25 m high deserve special mention. They are covered by grass and shrub vegetation. Small mounds and smoothly rolling plains are usually stabilized with grasses. Thanks to vegetation, hillocky sands contain small-grain soils in considerable quantities which favor formation of primitive, sandy desert soils. Some areas are composed of vegetated sand dunes formed under shrubs.

A specific eolian form in a desert is barkhan moving sand. Significant accumulations of sand in the form of barkhan chains are met with most often, and they resemble stiffened sea waves. Being mobile they present a certain danger for economic goals. Under a thick layer of barkhan sands archeologists dig out whole settlements with rich pieces of natural culture. Drifting sands especially caused great damage to the economies prior to 1917 when no works to combat them were conducted, but unwise land use led to new territories being affected. For instance, in the course of the operation of the Trans-Caspian railway each train was accompanied by a special team of workers who had to clear railway tracks of drifted sand. Disturbing news on the danger of drifting sands was published even in newspapers. By now the mechanism of formation and development of drifting sands had been studied well and methods for their stabilization and afforestation worked out. Nowadays these areas are not a threat, and moreover can be turned under proper management into agricultural lands.

2.2 Climate

The climate of deserts in Central Asia and elsewhere in the world is characterized by high air temperatures and a long dry period in the summertime. Although these features are peculiar to all deserts, they still vary from region to region. Geographical latitude and an altitude over a sea level influence climatic conditions of deserts. At the same time certain features characteristic of a desert climate are visible, too. Mean January temperatures in deserts of Central Asia vary from 4.1°C in the Northern Kyzylkum to 0.7° C in the Southern Karakum. Mean temperatures in July can be as high as 28°C in the north and 32°C in the south. Precipitations vary from 100-150 mm.

One typical feature of climate in the Central Asian deserts is seasonal precipitation. More than half of the precipitation falls in spring and the rest in late autumn and winter. In summer rains are very rare. So, there are two seasons in deserts: a dry season that lasts from mid-May through mid-October and a humid season in the remaining part of a year.

A small amount of precipitation combined with their seasonal distribution contribute to great air dryness, especially in spring and early autumn. High temperatures and considerable dryness of air throughout five or six months are responsible for high evaporation from soil and the surrounding atmosphere. This results in soil dryness, in drying out of the top sand layer and in the increase of transpiration of plants.

Because of mild, even warm winters vegetation of plants stops only for a ten day period. And in the Karakum and Kyzylkum nearly half of winters are mild or even warm. A long, vegetation period is characteristic of a desert climate, thus, a repeated blossoming of ephemerals and ephemeroids that are dormant in summer, and revive in autumn is quite possible. In oases, irrigation allows harvesting of two yields of early-maturing crops, or cultivation of such long-maturing crops as cotton, kenaf, rice and others. In the south – and this is very important – it is possible to grow fine-fiber cotton, one of the most heat-loving crops.

When there is an acute deficit of precipitation, droughts of various intensity and duration can occur. It is important to distinguish aridity and drought. Aridity is a permanent characteristic of a region, while drought is a temporary event that occurs under acute insufficiency of precipitation.

It is impossible to determine duration and persistence of such periods of the increased and insufficient moisture supply because of a short set of observations about atmospheric precipitation. Apart from long periods of increased moisture supply, an inherent feature of a changeable climate in Central Asia are the long, dry periods during which droughts can occur. Damage incurred by a drought to agriculture can be considerable. In dry years harvests on rainfed lands are reduced by 40%, and on irrigated lands 30%.

Wind is a noteworthy element of nature in a sandy desert because it changes the natural environment significantly. At a windforce of 4-5 m/s sand starts moving. In stronger winds a relief of sand gets altered, new barkhans appear and the form of old barkhans changes. Bare and poorly overgrown sands are drifting, the roots of plants become bare, seeds are blown out. Sand driven by wind covers engineering structures and wind blows out their foundations.

Strong winds cause dust storms that last in the Karakum and Kyzylkum for 20-40 days a year on the average. More often they happen in summer, then a wind speed reaches 10-14 m/s.

Frosts in spring and autumn are typical of the desert climate. During spring frosts plant, vegetation both on pastures and on irrigated lands is retarded, and a vegetation period gets shorter. This affects yields of desert and cultivated plants and the terms of their maturing. Pastures and irrigated lands alike suffer from frosts.

Hail, mists and thunderstorms are rare, atmospheric events in deserts. In northern regions during ten years of observations only two to five days with hail were recorded, and in the southern regions one to three days. Hail happens more often in spring. It damages plantings and gardens. In the Kyzylkum and Karakum mists are observed 10-20 days a year. Because of poor visibility, mists are very dangerous for planes and motor transport.

In general, deserts are characterized both by favorable and unfavorable climatic phenomena. They await more in-depth analysis. They should be also taken into consideration in the course of agricultural and industrial development of deserts.

A hot, dry wind called "*garmsil*" is very harmful for agriculture. These south-eastern winds blow with an average speed of 12 m/s. In some years *garmsils* can blow for two weeks. During these periods air temperatures can reach 43°C and a fertile soil layer can dry out completely. The effect of *garmsils* on plants is very harmful. During heavy *garmsils* even with copious water applications up to 80% of a yield can be lost.

Among local winds bringing droughts "*Afghanets*" is worthy of mention. This is a powerful rush of dry, tropical, air masses from the south that reach the Murghab and middle Amudarya oases. "*Afghanets*" is a very dry wind moving with great speed, and bringing much dust with it. The period of its action is shorter than *garmsils* and is equal to 2-3 days. It can blow up to 70 days a year. "*Afghanets*" is preceded by spouts and vortices, and comes from the south with a continuous wall of dust attended by a stormy western wind whose speed can reach 20 m/s and more. The wind depresses agricultural crops, covers fields with sand and dust. In early spring it brings rainstorms and a sharp drop of a temperature, even frosts. In winter it is sometimes attended by wet snow which leads to the deaths of cattle.

Pasture vegetation suffers very much from droughts. In dry years as a result of simultaneous action of soil and air, the grasses of pastures can dry out 10-15 days earlier and its yield decreases by 30 to 65%.

The climate of deserts also has unfavorable features that drastically change the conditions for development of an organic world, the lives of people, their economic activity. First of all, we should mention weather changeability in spring and autumn. Besides favorable seasons of a year, there is also a possibility of dry springs or cold springs with late frosts with delayed rains. Any of these conditions can lead to a yield failure of desert plants, especially of grasses.

Some winters can be frosty, with heavy snowfalls, snow storms which can damage animal husbandry seriously. In summer, very high temperatures are recorded; at times they can reach an absolute maximum of 50°C. Summer temperatures of 42-44°C are quite normal here, especially in the Karakum. In summer soils get heated during a day to 60-70°C, the highest temperatures being above 80°C.

High temperatures in summer and lack of precipitation lead to low relative air humidity. Unlike any other natural zone, air humidity in a daytime is not more than 30%, falling some days to 2-3%. In general, a snow cover in deserts of Central Asia is unsteady. But in some cold winters, snow lies for 35 days in the Karakum and up to 70 days in the Northern Kyzylkum. Sometimes its depth can be as large as 70 cm. Snow is especially dangerous at winds 4-9 m/s and frosts 20 to 30°C.

A specific wind regime is observed in the Central Asian deserts, and is connected with a baric relief and synoptic conditions. North-eastern winds are dominant here. Their average annual speed is 5 m/s in the west, 2.5 m/s in the south-east and 3 m/s in the east. The highest wind speeds are recorded in spring. In the Central Karakum strong winds blow ten days a year, in the Kyzylkum 11 days, on the Caspian coast 30 to 36, and in the South-Eastern Karakum up to 50 days a year.

Storm winds with a speed of 24-26 m/s are rare in deserts; they are recorded only one to three days a year. More often they are observed in the western part of Central Asia.

Apart from winds caused by a general circulation in the atmosphere there are local winds in various parts of a desert. Local winds, often strong and steadily blowing, are witnessed in the vicinity of mountain spurs and narrows.

2.3 Surface Water

Deserts are characterized by a severe water deficit. This is the only landscape zone which does not have surface water flow, where big rivers flow on the margins of a vast territory and do not receive any tributaries, while small and medium rivers end in a desert as blind deltas.

The biggest rivers in deserts of Central Asia are Amudarya and Syrdarya. Rushing from high mountains, they flow over a sandy plain to the Aral Sea creating a floodplain landscape. The Amudarya and Syrdarya rivers formed a *tugai* belt with specific soils, vegetation and animals. Part of their water percolates into the soil, thus, replenishing fresh groundwaters. Much water from the Amudarya and Syrdarya is withdrawn for irrigation. Not long ago the water of the Amudarya and Syrdarya was used for irrigation of fields in terraces above floodplains. Now, after construction of canals many kilometers long and inter-basin river flow transfers, the area of using these rivers became limited. Being transit rivers the Amudarya and Syrdarya do not contribute significantly to the transformation of adjoining territories.

Unlike the Amudarya and Syrdarya, the Murghab, Tedjen, Zarafshan and Kashkadarya rivers produce great influence on the territories adjoining their valleys and deltas.

Groundwaters are largely fed by seepage from rivers, main canals, irrigated fields, and temporary waterstreams. Precipitation that falls on a hard, poorly permeable clay surface (takyrs) is also a temporary source of fresh water. Water collected from a takyr watershed is an important source of water supply of desert-pasture, animal rearing and the water supply of population in a desert.

Humanity's impact on surface flow is revealed, first of all, by flow regulation and decreased run-off to seas and lakes. This happens due to the construction of many reservoirs which accumulate flood waters that were previously carried away unused to seas.

The extension of irrigated lands makes it necessary to increase water intake from rivers. Irrigation canals are built and water intake grows. This process leads, on the one hand, to complete disappearance of river flow into a sea, and, on the other hand, to the increase of groundwaters in irrigated zones.

Groundwaters occur in various soil horizons, therefore, pits are dug to different depths. In the Kyzylkum pits from 30-100 m deep prevail, in the Central Karakum from 10 30 m and in the South-Eastern Karakum from 100-200 m.

Groundwaters are largely saline. In some regions salinity is as high as 50-100 g of dry residue per one liter of water. Groundwater salinity in a desert differs from region to region. In the Central Karakum it is 3-10 g/l tending to increase at margins to 30 g/l, in the South-Eastern Karakum 1.5-3 g/l, and in some places to 10 g/l.

In the Kyzylkum groundwaters are less saline: in the north 3-10 g/l, in the south 5-10 g/l.

2.4 Traditional Artificial Water Reservoirs

More than 60% of groundwaters in Central Asian deserts are saline to different degrees. Fresh water is mostly in areas nearby deltas and river valleys, dry river beds, main canals, irrigated fields, and takyrs. People living in deserts learned to collect and store precipitation. They use the ancient, but reliable method of collecting precipitation from takyrs and takyr-like surfaces and their storage in shaft wells and covered reservoirs *sardabas.*

Long ago people extracted groundwater in piedmont deserts by construction of remarkable underground canals *kyariz* via which water was transported outside. A head of a kyariz usually is located directly at a mountain, while a mouth – via which water is supplied to the surface – is at some distance, sometimes dozens of kilometers, from a source.

If one gets down into a *kyariz* well, then through a narrow opening high above your head you can see sparkling stars, although on the surface it is day – so great is its depth. In a well you can smell earth, fungi, river freshness and the staleness of a cave. Under your feet water is bubbling, it flows down the walls and drips from a low dome.

Kyarizes are very sensitive to earthquakes that can both increase and decrease, or even completely stop a flow of water along *kyarizes*. For instance, after the 1929 earthquake a water flow in the Kazachy *kyariz* increased ten times, but this led to its destruction – it collapsed and was scoured.

Irrigation from kyarizes that was mentioned by historians as early as in 800 B.C. was practiced in many countries: Iraq, Yemen, Algeria, Morocco, and China. In Central Asia it was at one time widely used in Turkmenistan.

Whole generations of skilled craftsmen worked on the construction of *kyarizes.* The difficult and dangerous profession of searchers for subsurface water was passed from generation to generation, from fathers to children.

The wisdom of the people who constructed such superb, engineering structures still amazes us by knowledge of engineering skills and natural conditions of their territory. In recent years a method of groundwater extraction by drilled wells has become more widespread.

2.5 Soils

Soils in deserts are very thin, humus content in them is meager. They do not possess any visible alkalinity and texture, and they are saline nearly everywhere.

Soil-forming processes in deserts of Central Asia develop under specific temperature and water regimes that get established after long, intensive heating of a soil surface in summer and its cooling in winter, air dryness, a low content of humus, etc. The nature of parent rocks is essential for soil formation. Soil types over vast expanses of plains are monotonous, because soil varieties develop under very similar lithological and extra-arid climatic conditions.

On a map of soil geographical zoning the Turanian plains are referred to as a desert zone composed of gray-brown and sandy desert soils. The vastness of the territory and, as a result, the presence of different natural landscapes contributed to the formation of local features of the soil cover. The main soil types here are sandy desert, gray-brown, takyr-like, takyrs and solonchaks.

Sandy desert soils develop on surfaces overgrown with grass vegetation. The development degree of a profile of sandy desert soils is different in various parts. In hollows and lower parts of slopes of ridges and mounds covered with dense grasses, profiles are well developed, whereas in areas less stabilized with vegetation a profile of these soils is practically not differentiated into horizons.

In conditions of arid climate, the sandy deposits are overlain with low-humus, mostly non-saline or slightly saline soils. Aggregation of sandy soils is poor. Some compacting is witnessed to a depth of 20-30 cm.

A top (2-5 cm) layer of typical sandy soils is a loose and permeated by root necks of ilak. In localities where epelek (military grass) prevails in the grass growths that are often observed in the Karakum and Kyzylkum, a sandy layer is not significant as roots of this cereal start developing directly from the surface. The next horizon to a depth of 10-15 cm is compact, poorly textured with many roots and to a depth of 30-50 cm there is a layer with some vertical fissures. Sometimes at the foot of this layer compacting with fine disperse material and carbonate is found.

These soils are composed of sands in which fine-grain particles prevail that make up 50-60% of the whole fine-grained mass. A silty fraction in the top half-meter layer averages up to 4-6% and occasionally 8-9%. Some increase of its content is registered in a layer of non-projective layer of grass.

Sandy desert soils are hardly saline because the content of solid residue is estimated as hundredths fractions and chlorine as thousandths fractions of percent. The level of carbonates is not high – 3 to 5%, some increase of their content is traced in the bottom of the top, half-meter layer. Gypsum is seldom present in a profile of these soils, and the humus content in sandy soils is low. Even in the most biologically, active parts which are densely permeated by roots, it is not more than 0.3-0.5%.

For a long time many scientists claimed a possibility of soil formation processes in eolian sands. The dominating opinion was that these sands are constantly moving, and that is why no features of soil formation processes can be detected on them.

Meanwhile researches of many scientists have revealed on sands some specific, poorly developed, desert soils having features inherent only to themselves. Because of the insignificant brokenness of relief, low fertility and lack of water sources these soils are unsuitable for farming. But in some regions of Central Asia, sandy desert soils are used for the cultivation of melon crops, alfalfa, and cotton. In some areas local populations have long practiced melon growing on sands.

Grey brown soils are automorphous desert soils. They are developed over a vast territory on plateau-like, residual hills. They are formed, largely, on eluvium of parent rocks of the Tertiary plateau, as well as on the ancient proluvium of piedmonts.

Soil forming rocks are mostly loamy and silty, small-grain ground usually containing inclusions of crushed stone forming phenocrysts and occurring on the soil surface. The thickness of the small-grain, ground layer varies from 50-200 cm.

Morphologically the gray-brown soils are homogeneous. On the top a honeycomb porous, non-compacted, light grey crust (2-5 cm) can be seen; then comes a finely banded, loose layer (3 to 6 cm) passing into a brown, compact, lump-cluster, with solonetz features horizon (10-20 cm); this in turn is underlain by a less compact, brown layer (10-20 cm) transitional to the parent rock. Inclusions of crushed stone, the amount of which increases with the depth, can be traced on the surface and along a profile.

Generally, the top layer of these soils to a depth of 20-40 cm is slightly saline and in some spots is desalinized. These soils, like any other automorphic soils, have a low humus content. This is a result of intensive mineralisation of organic matter in the soil and an arid climate. In the middle, most wetted part of the profile, one can see some gleyisation and increase of an exchange capacity due to more intensive weathering of sediments. At the same depth brown coloring of the profile becomes more intensive.

A level of water soluble walts in these soils is mostly insignificant, and is equal to 0.5%. In the lower part of the profile at a depth of 25 or 35 cm the content of salts starts increasing to 1-2%. At this depth fine crystalline outcrops of gypsum usually appear which pass downwards into a continuous gypsum layer in parent rocks. The gypsum content in such cases is seldom more than 50%. A carbonate content in gray-brown soils is the highest in the top of the profile which is related to their desert origin. Carbonate content is high and reaches 8-16% (in CO_2).

Humus content is these soils is low, more often 0.5-0.7%, occasionally 1-1.2%. Related to humus the content of total nitrogen is also insignificant at 0.03-0.05%.

Takyr soils are developed over relatively young deposits, in ancient deltas, upper river terraces and piedmont inclined plains composed of alluvial and proluvial deposits.

The surface of these soils is crossed with fine fissures making a polygonal pattern. On top a slightly compacted, rare-porous crust (2-4 cm) is seen, then comes a brown, non-compact, thinly layered thinly banding, finely banded horizon (5-8 cm), and a lighter colored lumpy saline layer (10-20 cm) directly underlain by the parent rocks.

These soils vary in salinity, and everywhere. On piedmont plains they are more saline than in ancient deltas. A solid residue within a one-meter series varies from 0.3 to 1%. Carbonate content here is high – 7-15%. Total alkalinity in top horizons is as high as 0.04-0.07%. The gypsum level is not high at 0.5-2%. A humus content in these soils is low, about 0.5-0.8%, seldom up to 1%. A content of total nitrogen is low too, 0.05-0.1%.

Development of takyr soils is dependent on surface wetting of saline, clay-loamy deposits under desert climatic conditions. Precipitation falling in autumn, winter and spring wets takyrs to a depth of 20-30 cm. Quick saturation of the top layer leads to the formation of surface flow that temporarily accumulates in lowerings.

A texture of deposits making up takyrs is largely clayey or loamy. The content of physical clay reaches 50-80%, silty particles up to 20-30%. An exchange capacity is 6-11 mg per 100 g of soil.

Nearly all takyrs are related to high solonchak, medium and highly saline soils. Throughout the whole profile of takyrs considerable concentrations of water soluble salts are detected. Top horizons feature low salinity rarely. In general, a content of solid residue varies from 0.5 to 2%. Some alkalinisation of the surface layer of a crust is typical. A humus content in takyrs is about 0.5%.

Salinity, a low humus content and very unfavorable, water physical properties prevent takyrs from being used for agricultural. Development of these soils for growing agricultural crops requires radical, land reclamation.

2.6 Vegetation

Vegetation of desert territories is typically xerophytic with a great number of endemic varieties and genuses.

It seems that in no other natural zone on the globe is a dependence of vegetation on geographical environment so clear-cut as in deserts. First of all, much desert vegetation is well adapted to extreme conditions.

Physico-geographical processes that change the Earth's surface, its climatic, water and soil conditions greatly affected the composition and distribution of plants, caused a change of varieties. Anthropogenic processes also influenced the desert flora. Excessive heat moisture deficit in summer, an often cold winter, a short spring and autumn, salinity of soil and groundwaters, sand drifting – all these are responsible for specific features of desert plants and their adaptability to the ecological conditions.

Sands are distinguished by most favorable natural conditions: they feature high filtration and low water lifting properties. Moisture, accumulated by sand in the mesothermal period, evaporates only from top layers (down to 100 cm). At the same time featuring high porosity they are characterized by relatively low, water intake capacity and a low wilting coefficient.

Communities of psammophytes on low in humus, loose, often mobile ground are typical of sand regions. The main representatives of tree-shrub vegetation are *Haloxylon aphyllum, H. persicum, Salsola richteri, Calligonum spec., Ephedra strolilacea, Halathamnus spec., Ammodendron conollyi, Astragalus spec.* and others. Grasses are represented by *Astragalus transcaspicus, Stipagrostis spec., Anisantha heliotropicum, Eremopyrum spec., Argusia spec., Agriophyllum spec.* and others.

The most widespread representatives of tree-shrub vegetation in sandy deserts are *Haloxylon aphyllum, H. persicum. H. aphyllum* mostly grows on thick, ridge and heavily broken, barkhan mound sands, usually occupying positive relief forms (tops, slopes). A clear-cut replacement of associations is observed with changes of the relief elements.

H. persicum (Plate 2.1.) is typical mostly of lowered relief forms where sands are more fine and saline. *H. persicum* is thicker than *H. aphyllum*; its stems are more fleshy, of a darker coloring and drooping. In *Haloxylon persicum* thickets its density averages 400-500 trees per 1 ha. A height of some old specimens reaches 6-7 m, and they weigh up to 1 ton. The weight of the total mass is as high as 40 t/ha.

In the vegetation of sandy deserts of Central Asia, grasses make up a significant share. They also get adapted to growing on different soils. Grass plants, the so-called ephemeroids and ephemerals, can be 5-10 cm high and have a short root system. They have a very short, vegetation period, but develop very quickly. They manage to complete their life cycle from germination to ripening of seeds and dying during 1-1.5 months in spring when rains fall. Ephemeroids are perennial plants. In summer they dry out but keep their root system alive, while in autumn when rains start falling they germinate again. They include *Ferula assa foetida, Dorema sabulosum*, some *Astragalus spec., Carex physodes, Poa spec.* and *Tulipa spec.*

Carex physodes is a very valuable forage grasses. It propagates both vegetatively and by seeds. Having a good root system, it fixes sands well. *Carex physodes* growth depends on rainfalls. In an averagely wet year its dry mass is 13-25% less than in a year with rainfalls higher than in an average year. Its biological properties help it to maintain constant density and, "the annual yield of feed mass." Therefore, even in dry years *Carex* grasslands provide more feed than others. It should also be stressed that *Carex* grasslands are more resistant to grazing. Significant loosening of ground by cattle, covering of its overgrown part with earth do not affect it so severely as other forage grasses.

Apart from the above advantages *Carex physodes* features high feeding capacity. During flowering its content of protein is 34% and fat 20% more than in alfalfa. Making at least a third of the forage mass eaten by cattle it becomes the principal element of the rangeland diet for cattle. Surviving well in summer in a dry state, *Carex physodes* can be a feed in different seasons. Sheep and lambs prefer eating it in spring, but in summer, when annual grasses are wilted, and wormwood and thistle are still astringent and bitter, sheep avoid them.

Ephemerals are annual plants. Their life cycle is the same, but they are met among grasses in spring and early summer. In years with insufficient rainfall in spring ephemerals can wither out without even producing seeds. This affects a composition of desert vegetation. Deserts also support vegetation that adapted itself to saline soils. These are halophytes – *Arthrocemum spec.*, *Salsola spec.* and others.

Soils and vegetation of deserts are characterized by a complex and patchy pattern of distribution, which is associated with close occurrence of different forms of relief and soils differing by a water content and structure. Depending on the interaction of various natural conditions in Central Asian deserts, certain types of deserts and their varieties have been formed here. The emergence of these deserts is a result not only of a specific combination of natural components forming different phytocenoses, but of the impact of people too.

It is interesting that some plant formations grow in certain, genetically specific localities. *Haloxylon-Carex physodes* and *Haloxylon-Salsola richteri-Carex physodes* formations in which *Stipagrostis karelinii* is often met prevail on overgrown sands formed as a result of deflation of ancient, alluvial deposits. In inter-ridge hollows, *Artemisia-Salsola* formations dominate. *Haloxylon aphyllum-Calligonum spec.-Stipagrostis karelinii* formations are typical of barkhan sands.

On sands formed from marine Khvalyn deposits, the *Calligonum spec.-Salsola richteri-Stipagrostis karelinii* formation dominates. Rare shrubs of *Haloxylon spec.* and *Agriophyllum spec.* are met here, too. Thickets of annual *Salsola* grow in localities with more saline soils. Sands overlying piedmont, alluvial-proluvial deposits are grown largely with *Calligonum-Salsola spec.-Carex physodes* and *Carex physodes-Calligonum spec.-Anisantha Heliotropicum* formations. In some places where sands are thin, *Artemisia* spec. is met. For broken and semi-overgrown sands *Stipagrostis spec.*, *Agriophyllum spec.*, rare shrubs of *Fabaceae Calligonum spec.* are typical. In places where groundwaters reach the surface, thickets of annual *Salsola spec. and Alhagi persarum* are met.

Takyr-like, gray-brown soils developed among ridge sands are overgrown with *Carex pachystylis- Bulbosa spec.* formations.

Psammophytes are met on Pliocene sands. Dominating here are *Haloxylon aphyllum*-Carex *physodes* and *Stipagrostis karelinii* formations that usually grow in positive relief forms. *Calligonum-Salsola richteri*-ephemeral vegetation occurs in the bottom of ridge slopes and in inter-ridge hollows where thin sand sediments and ground with a higher moisture content are met. *Artemisia spec.*, *Astragalus spec., Convolvulus subchirsutus, Salsola orientalis* and others grow here, too.

Nearly all areas covered with different varieties of *Artemisia* are used for year-round grazing of sheep and camels.

3. People in Deserts

A.G. Babaev, A.A. Arnageldyev

ABSTRACT: Deserts were in ancient times the basis of human civilization e.g. in Mesopotamia. Humankind in Central Asia from the earliest days resourcefully developed sophisticated ways of coping with the extreme conditions. The chapter explains how and where people lived: they learned to irrigate lands; they stored food, bred cattle (sheep, camels etc.), and built living quarters suitable for weather extremes. Modern technology has helped create desertification, but also provided new methods of coping with the environment. Science based practices must be used to maximize conservation while developing the resources of the desert.

Nature is the basis of the material life of the human species, one of the critical conditions for the existence of people. By improving productive forces, people more actively influence the natural environment, change it, find new resources and other combinations in it, appraise anew the earlier known features and quality of nature.

Scientists have shown interest for a long time in the mechanism of interaction in the "people-environment" system. This problem is old, complex and versatile, and extensive materials and experience have been amassed on it. Methods and approaches to harmonization of interaction between nature and society are very different.

At present somebody possessing a powerful technology can transform landscapes of vast territories, and interfere with the course of some natural processes. And the more that human society interferes with the natural environment and uses its resources, the greater is the impact on natural ecosystems. As a result, voluntarily or involuntarily, the dynamic equilibrium of nature is broken, and the living conditions of people become worse.

The growing number of ecological crises are closely connected with the social, economic and political structure of the present-day world. And it is not accidental that in this complicated and dangerous situation most vulnerable are economically underdeveloped and politically unstable countries of Africa, and, above all, in the Sahel zone. That is why socio-economic development and the possibility of state planning and regulation become critical factors for the success of nature conservation and desertification control activities.

It is well known that deserts always were and are still difficult for living. But notwithstanding this fact it is in desert landscapes that highly-developed ancient civilizations appeared: Egypt on the Nile, Mesopotamia in the interfluve of the Tigris and Euphrates, ancient Khorezm in the Amudarua delta, ancient Merw in the Murghab delta, and Sogd in the valley and delta of the Zarafshan. The critical factors that attracted people and formed the basis for births of civilizations were most importantly: a hot, dry climate, long and warm season(s), fertile soils, and rivers that do not dry out in summer. But above all it was the labor of many generations of people which created these civilizations.

People by using natural resources, make the most essential and visible transformations in the desert environment. The more intensive the search for and withdrawal of these resources are, the greater are the changes in the natural environment. Excessive use of natural resources of deserts can change the original landscape beyond recognition. Deserts are pastures which have been turned by overgrazing of sheep into bare sands, irrigated lands under excessive water application into solonchaks, while wise water application can transform them into an oasis. They differ from each other significantly, although initially they made up a single, desert landscape. Therefore, overgrown and mobile sands, natural and artificial pastures, irrigated lands and an oasis can be justly referred to as a desert zone.

Deserts are very demanding to those who start developing them. Unwise management of the desert environment, disregard or ignorance of its specific natural processes invariably lead to negative implications.

In deserts inattention for the laws of nature has quicker and stronger results that change the natural environment so that its economic management becomes impossible or too complicated. On the other hand, wise activities based on scientific knowledge and people's experience give positive results, and enable management of desert resources without damaging the desert environment significantly.

For many centuries those living or associated with deserts were guided by everyday experience, observations and sometimes intuition. They learned the laws of natural processes in practice, and often they could not explain them properly. Nowadays the use of a desert's natural environment has a scientific and engineering basis. Interactions with nature proceed from scientific cognition and explanation of the action of a natural mechanism and physiographical processes. Through science the people are learning how to manage the geographical and environmental processes, to cope with negative events, and often to know beforehand their likely aftereffects. Very high temperatures in summer, insufficient rainfalls, a lack of fresh water, wide occurrence of saline groundwaters, limited opportunities for farming development, very unstable weather dynamics in spring and winter, sparse vegetation, drifting sands, frequent sandstorms make life in a desert very difficult.

Absence of roads, and even impassability in a dry season that lasts more than six months are typical of sandy deserts. This makes people dependent on means of transport, limits their choice, speed and carrying capacity. But nevertheless a desert with its hot and arid climate was one of the ancient human habitats, a place where ancient civilizations developed. Some peoples roamed in deserts with their cattle, others settled permanently in oases. Egypt between the Sahara and Arabian deserts, Mesopotamia nearby the Syrian desert, Khorezm between the Karakum and Kyzylkum deserts, ancient Merw between the Eastern and Southern Karakum, Sogd between the Kyzylkum desert and the Golodnaya steppe – these are only some of the ancient civilizations that left a significant imprint in science, technique, culture and everyday life of humanity.

3.1 Agriculture on Irrigated Land

Of course, it was not the desert itself that gave birth to ancient civilizations. They were created by the labor of many generations of people around water sources – in valleys or deltas of rivers, such as the Nile, Tigris, Euphrates, Amudarya, Syrdarya, Hwang-Ho, Murghab, Zarafshan and others, on piedmont plains irrigated from small rivers. For instance, excavations on the plain nearby Kopet-Dag at Annau settlement have revealed that this place as early as in the 5th millennium B.C. was populated by ancient farmers. In the third century A.D. in Nisa nearby the present-day settlement Bagir there was a residence of Parfenon tsars. Ancient centers of civilizations were formed alongside ancient trade roads, at junctions of desert and river valleys, a desert and piedmont plains.

If there was enough fresh water, one could enjoy such favorable aspects of a desert climate as long vegetation and frost-free periods, a high sum of positive temperatures in a year. All the above made it possible to grow valuable, southern, heat-loving foodstuffs and crops usable in craft work e.g. textiles. Possessing irrigation water, people could alleviate the unfavorable influence of scorching heat, dry or severe winds for plants. In general, irrigated farming is typical of arid territories. It demands well-regulated use of water, timely and adequate supply of water to fields. Here irrigation is needed: canals with design inclination, dams and water bodies. It was necessary to repair and maintain all structures operationally. Thus, the art of hydraulic engineering was developed. In some localities irrigated farming remains small-scale and rather primitive, in others it covers vast areas and is technically advanced (Plate 3.1.).

At the same time all works on construction, maintenance and repair of irrigation systems became centralized. The centralized management of the society became more effective, too.

Irrigated farming needed not only water, but advance knowledge of the time of river floods, the size of the expected river flow to fill water reservoirs and to avoid inundations and destruction of dams. This gave rise to the development of exact sciences, mathematics, and astronomy. And this led also to social differentiation of a society.

In ancient Merw a graduated board was installed in a special water reservoir that was filled in spring with water from the Murghab. It was used to determine the coming of a high-yield or dry year. And water was distributed in accordance with a forecast. In this way people living in oases gradually, over many years, learned how to transform deserts into a cultural landscape, to control water use, to develop farming, crafts and trade.

At different stages of social and technical progress, people could not use water from a big and small source with equal ease. Water management of a big river demanded great material and labor inputs, engineering facilities and centralized management. A small source of water (an interim waterstream, a water spring, a small river) can be possessed by a relatively small group of people; that is why it is not accidental that the earliest settlements of farmers practicing artificial irrigation were found in piedmont plains. Small rivers here have some natural advantages i.e. they do not have vigorous spring floods; coming out from narrows to plains, these waters flow more quietly in not very wide river beds, feeding from springs that do not dry out in summer. Piedmont plains are distinguished by fertile soils and convenience of irrigation: it is not necessary to build the main canal – a river bed can be used as such; it is not necessary to dig inclined trenches by gradually deepening the irrigation canals because a piedmont plain has natural sloping that guarantees water supply to fields by gravity; it is not necessary to carry out significant leveling of irrigated fields because the bottom of piedmonts is rather well leveled naturally, and natural drainability prevents soil salinization, etc.

Under such natural conditions, irrigation development of desert lands was possible even with the primitive tools of a primeval community.

Later in the slave-owning and feudal societies, irrigation was developed largely in deltas of rivers with medium flow. Water flows quietly in them, delta branches are used as main canals, while vast areas fertilized by silt deposited from river water can support large populations. That is why Annau, Nisa appeared earlier than ancient Merw, Daghistan and made way for them as being the larger and better organized communities on the Murghab and Atrek rivers. The largest states, such as Khorezm in the Amudarya delta, Bukhara in the Zarafshan delta, appeared near large areas of water when conditions became conducive to formation of centralized states.

Piedmont plains and river deltas were chosen for living not only because of the convenient management of water of small rivers and delta branches, but also diversity of the natural environment. Piedmont plains are located at an interface of mountains and deserts, while deltas are surrounded by a desert. This facilitates public and territorial division of labor. Diversity of natural conditions provides for exchange inside a society and with neighbors, between cattle breeders and farmers in a desert.

3.2 Nomadism

Unlike settled farmers, the cattle breeders roamed a desert gradually changing its natural environment. While rearing sheep and camels, nomads had to move with their cattle from one pasture to the other, from one water pit to the other. While on seasonal migrations some of them lived in temporary seasonal settlements from which they drove their cattle to distant pastures and returned. The size of settlements, the times of pasture usage, the distance of cattle driving depended not only on pasture feeding capacity, but also quantity and quality of water at the disposal of nomads.

Both farmers and nomads mastered the arts but those arts that make their living in a desert easier. So that they can live in a desert, over many centuries the people living in the Karakum and Kyzylkum deserts have thought of clever systems of people's hydraulic construction and agricultural practices. They mastered the selection of domestic animals and various crafts.

Nomads knew well their natural environment, and could find water in such places where it should not be by all indications; ploughed small plots with limited water application and even without it; dug pits unique for their depth, design or method of filling with water.

People living in deserts have specific diet and feeding regime. Thus, people living in the Karakum and Kyzylkum deserts eat mostly meat and avoid fish. They thought of original methods to store meat in a hot climate – they fry mutton chopped to small pieces, melt down lard, mix it with the chopped mutton, put this all into a big, enamel pot. After the meat and lard has solidified, the *kaurma* is ready. The product thus made can be stored for a long time while retaining its pleasant taste. From *kaurma* one can easily and quickly make a *bachansky shurpa* (a kind of a soup) by placing the required quantity of *kaurma* in a mug and filling it with hot tea. It is possible to cook roast meat having heated *kaurma* on a pan, or any other meal.

People living in deserts are well acquainted with different diseases, including those that are spread by rodents and pests. Thus, in old times in Central Asia people suffered from eye diseases caused not only by a bright, blinding sun, but to no less a degree by smoke that filled nomad tents (*yurta*) in winter. But desert people intelligently avoided gastric illnesses because they drank only boiled water. They drank hot tea to quench their thirst and to lower body temperature, and this made them accustomed to boiled water. And even now this is still common practice among all peoples living in Central Asian deserts.

We can give here many more examples how people adjusted themselves to specific conditions of deserts. But they all have one thing in common: knowledge of laws of desert development helped people apply the most reliable methods and ways of living and working in the natural environment of arid zones.

The main occupation and path of desert development was cattle breeding: rearing of sheep and camels. Very likely there had been hunting earlier and then parallel with cattle breeding, but this had long ago ceased to be a means of sustenance in deserts.

Nomadic cattle breeding by itself is the simplest and easiest occupation, and a reliable source of existence in a desert. But it is based not simply on picking of feed by animals, but, above all, on the great experience of nomads, their deep knowledge and understanding of the natural environment where they lived. The choice of sheep and camels for rearing is in itself a proof of how well they knew desert conditions. Sheep and camels were best adapted to a hot climate and saline water, to a specific set of forage plants that change their nourishing capacity, mass and edibility from season to season. Moreover, the annual biological cycle of sheep correlates in general to a cyclic development of nature in a desert. For instance, in Central Asian deserts sheep give birth in spring (March) when many green plants appear. As a result lambs soon change over to green feed, and in June when it is very hot and grasses are withered they can already feed on branches and stems of a tree-shrub vegetation.

Thanks to adaptation of sheep and camels to extreme desert conditions and their high, economic value for many centuries, these animals go together with humankind in a desert. Of course, the breaking of this cycle – in some years it happens due to a dry or over long spring, or due to a dry or too lengthy autumn – causes the death of many sheep, thus, reducing their flocks. One of the main riches of the Central Asian deserts is Karakul sheep which were bred and now are reared on plains in a hot and dry climate, feeding on specific forage plants.

Sheep are valuable animals indispensable for a desert. Sheep wool is used for making carpets, sacks, felt (*koshma*), ropes and many other things necessary in everyday life. Felt or *koshma* is used to cover a wooden frame of a *yurta* (nomadic tent), or a floor. Sheepskin is used for making fur-coats, fur caps; sheepskin is dressed to make footwear, saddles.

Camels are very important for life in a desert, too. They also provide meat, wool and skin. Camel skin is used for making leather buckets to bring water from wells. For 18 months, milking female camels give milk which exceeds cow milk in fat content. Diluted with water and leavened, it turns into a vitamin, cultured milk product indispensable in hot weather. The Turkmens call it *chal*, and it soothes thirst and hunger. In a desert *chal* is an additional source of water, too.

A camel is also a working animal. Camels help getting water from pits, when moving to some other place camels carry *yurtas,* household things and people. Thanks to the special structure of its hoof, a camel is well adapted to strolling over sands. It is very robust and not difficult to water and feed. Being of universal use a camel is indispensable both for cattle-breeders and farmers. Camels do not need grazing. Collecting feed from a vast territory, camels wander far off in a desert and graze alone or in small groups. In search of feed they are capable of covering great distances. But no matter how far they stray, after several days have passed, those camels return to their pit for drinking.

Sheep and camels support subsistence or semi-subsistence economy of nomads. What nomads do not have are rice, wheat and vegetables. Flour for baking bread is supplied from an irrigated zone. Barter relations are established between farmers and nomads on the basis of a geographical division of labor. But requirements in vegetables are not met to the full, although their deficit is partially covered by melon crops that are grown on small areas in a desert.

People living in Central Asian deserts have bred a shepherd's dog - *alabai* (a hybrid of a sheep dog and taaza), a hound and hunting dog. The *Alabai*, a large, fierce dog but obedient to its owner has remained till now the main "helping hand" of a shepherd. The dog helps him manage the flock while moving to another place, changing direction or camping. The dog can find a sheep that strayed from a flock and drive it back, can fight a wolf, and endure heat easily.

3.3 Housing and Living

Mostly people in a desert live in *yurtas*. Different peoples can decorate them differently, but the design of *yurtas* is nearly always the same. A wooden framework is lined on the inside to a half height with reed mats and on the outside it is covered with *koshma* (sheet felt). On top they put a wheel (*tuinik*) covered with *koshma* for ventilation of the indoor space. Such a *yurta* best suits the nomads roaming in a desert: it is cool inside in summer, warm in winter, it can be quickly installed and dismantled. The shape of the *yurta* is such that sand winds can freely flow round it, and no sand mounds accumulate nearby. Nowadays the *yurta* is a convenient "house" in a desert.

In industrial settlements that appear in a desert, they mostly build modern, urban 2- or 3-storeyed houses made up of prefabricated blocks. Their production, transportation and installation are less costly and quicker than construction of mud-wall houses.

Central estates of cattle-breeding farms include mostly standard shield houses, plastered inside and outside, with a gable roof. *Yurtas* are used now only by shepherds when they drive cattle to summer and winter pastures.

They successfully cope with such problems as water supply, air cooling indoors in summer, control of sand drift, and the application of achievements in science and technology. Modern means of communication e.g. motor and air transport, radio, TV etc improve the living standards of rural populations bringing them closer to that of urban populations. But, in general, the ancient way of life of people in a desert is still maintained. In typically cattle-breeding farms it is preserved better than in industrial and farming centers where engineering facilities are applied wider.

People's activity in an arid zone interferes with the natural ecosystem often incurring damage to the society itself. Of course, everything depends on the intensity of the use of nature which, as is known, increases with every new stage of social development. Nowadays it is sometimes even necessary to introduce planned ecological management, to ameliorate damage caused to the natural environment, to preserve some natural resources and to compensate for excessive loss of others.

Rational nature management, revival, even improvement of the natural environment, and the application of a complex of reclamation measures are especially necessary for the deserts, because upsetting the equilibrium in natural complexes of the arid zone often causes irreparable damage. Damage incurred to a desert can limit and even put an end to economic activity on a vast territory.

Humankind can enhance the role of individual components of the natural complex of a desert. Such a component might be fresh water. Therefore, in a desert consideration of any economic issue starts with estimation of available water resources. In areas where it becomes possible to increase their reserves the natural complex gets drastically changed: it is characterized by a new quantitative and qualitative correlation of components, and their interaction. Those who possess water can purposefully influence the natural complex.

In a desert, water is used to alleviate the effect of excessive heat and air aridity. Around houses trees are planted that give shadow, fields are cultivated to abate air dryness and to get food. Such activity not only transforms nature, but also changes the economic and everyday way of life. Water directly influences a population density in a region, localization of population and its economic activities. This is mostly well seen on irrigated pastures, irrigated fields, in production areas.

Social and economic interests urged people to migrate to a desert that attracted people by its resources the use of which demanded much effort from a farmer and a cattle breeder. In an irrigated zone people can grow two, and in the southernmost areas, even three crops of fast-maturing plants. In the Karakum and Kyzylkum deserts cattle grazes all year round, something which is impossible in other natural zones. Everywhere it is necessary to provide cattle with warm premises and forage for winter which increases costs of cattle growing. In a desert a usually mild and short winter, and availability of grass on pastures make rearing of sheep and camels less costly. Apart from pastures people found other necessary natural resources in a desert. A tree-shrub vegetation gives them fuel, logs to fix walls of pits; other plants give stable natural pigments to dye sheep wool; others provide tanning substances or means for linen washing, still others are used for disease treatment.

But not all the natural riches of a desert stood the test of time. Some of them (dye stuff, tanners) have lost their competition with modern, chemical substances, although natural, non-fading pigments are still used in carpet-making. Walls of pits are now fixed with reinforced concrete blocks.

3.4
Modern Life

With the development of new socio-economic conditions, the people-desert relationships have changed drastically. These new relations are best seen in the involvement of national specialists not only in industry and agriculture, but in their management, as well as in communication (transport, radio, TV and press), in new kinds of raw material and fuel produced in a desert, in up-to-date technical facilities, production processes and products of labor. New attitude to deserts are also revealed in the changed outlook on the management of nature: the use of resources should go together with an insight into the future, with compensation for the damage incurred to the environment, with maintaining an optimal correlation of components in natural complexes.

At present the significance of deserts as a human habitat is changing. The number of nomads is gradually diminishing. Small settlements of cattle-breeders have been replaced by larger settlements and farms. Their appearance and economic functions have changed. At the same time, the percentage of urban populations engaged in fuel, chemical, construction and service industries is growing.

Nowadays settling is often associated with natural resources production, the construction of major irrigation canals, oil-, gas- and water pipelines, and creation of large cattle-breeding farms. Specifics of industrial enterprises, irrigated farming, grazing cattle breeding have a clear-cut effect on types and population of settlements, and their location often takes into consideration production functions of this or that settlement.

New prerequisites for settlement in a desert are not always combined with favorable, water-climatic conditions. Even in a desert there are areas with the most severe, natural conditions where dust storms and strong winds occur most often; areas with highly saline soils and groundwaters, and sparse vegetation. Earlier people avoided these areas because it was very difficult to rear cattle there. But today when oil or gas reserves or other natural riches are found in these areas, then new settlements appear nearby. In this case people are provided with all the necessities and those who work there get a considerable bonus to their salaries.

Modern forms of transport (cross-country vehicles, planes and helicopters) have changed life in a desert: distances become shorter, there are no more places in a desert that are inaccessible or accessible only with difficulty because of their remoteness, lack of roads or fresh water supplies. Moving with cattle to a new plaster, shepherds send their families, *yurtas* and personal belongings by cars.

A desert tests people rigorously. People work on oil or gas fields, pastures, irrigated lands, transport in open air where a daytime temperature in summer may reach 40-43°C. Under severe heat people become languid, move slowly; labor productivity drops by over half. In a desert the ecological situation itself suggests to people the best way of life. Here everything is subject to the task of self-isolation from heat, lowering an air temperature in places where people live, work, and rest. And these facilities have been much improved. Modern urban buildings in oases and in a desert nearby production sites are also provided with protection facilities so that desert dwellers avoid as much as possible the direct sun rays, and find themselves in a healthier microclimate secured by orientation of windows, application of heat-insulating materials, shading of houses by vegetation, air conditioning, etc.

In any natural zone, human life is dependent on natural and economic conditions. In a desert this dependence is intense. But with progress in social and economic spheres, improvement of technique, achievements in industry their dependence on natural conditions becomes weaker, more indirect, but does not disappear completely.

People who are involved in cattle breeding have the most solid and direct ties with nature. Seasonal movements of cattle were and still are caused by seasonal changes of forage and drinking water resources. Unavoidable and numerous journeys to new places, and long trips influence, in their turn, many aspects of the life and labor of cattle-breeders and family. Constant movement limits their demands in comparison to settled farmers, inbreeds in them certain production skills, traditions, customs. Parts of these have survived to the present, other parts have disappeared or been transformed under new social conditions. The old, disappearing customs are replaced by new, more up-to-date ones, meeting to a greater degree the spirit of the age.

Achievements in science and technology gradually influence life in a desert e.g air conditioners appeared for public and personal use which lower an air temperature in public buildings and living quarters. Delivery of water is widely practiced to pastures where there are no water sources. Apart from phytoreclamation they apply chemical and technical means of drifting sand fixation, build artificial water catchment sites where there are no clay takyrs. By improving vegetation cover, planting large shrubs in places where it was destroyed, seasonal pastures are turned into perennial pastures. Such continual pastures make further driving of cattle unnecessary. Mechanized delivery of water from pits makes camels as a pulling force redundant. These are examples of new interrelations in a "human-desert" system that are impossible in nomad life.

Interaction between humankind and deserts have varied in the course of history. This leads to a different dependence on nature, a different attitude to it, in particular to a desert. The deeper a science delves into laws that control life in a desert, the clearer are the changes that go on there, their causes and awaited results.

Relying upon progress in science and technology, people uncover the innermost secrets of nature, learn about the many laws that regulate it, and so the study and development of nature is scientifically based. At present we speak not about harmonious relations between society and nature, but instead about wise environmental management, the protection and purposeful transformation of nature and even about control of natural processes, prediction of their course of development and consequences. Rational eco-management should consider not only the needs of a present-day society, but future interests, too. It is necessary to put into practice the achievements in science and technology, to minimize the possibility of undesirable consequences, to compensate for damage incurred to the nature. As a result behavior in a desert must be based on a thoroughly developed scientific theory. This is necessary to assess what has been achieved in desert development, and to reliably predict what changes can be awaited in the natural environment and the lives of people in the arid zone in time, space and in quantitative and qualitative terms.

4. Population and Settlement

A.B. Batyrov

ABSTRACT: Central Asia has a very unusual, ethnic population structure marked by a comparative lack of tension. For each republic in turn the geography, history and sociology of the inhabitants in general is elucidated. The reasons for the present rapid growth both of the total numbers and the rural population are explained, offset by the possible emigration of Russian communities whose role has been affected especially in this century by unique historical processes. Urban and rural settlements are portrayed, contrasted and placed in the context of inhabiting the deserts especially in small oases.

As of 1 January 1996, the population of Central Asia was 54.4 million or an estimated 1% of the world population. The average population density was 13 per. km^2.

Central Asia is an unusual region in respect to settlement of the population, specifics and variety of settlements. Here, unlike many European countries, the rural population is large in percentage terms, and the citizens are distributed very unevenly which is associated with regional specifics of historical, natural and socio-economic conditions.

The results of the researches on social-economic problems carried out by the Desert Institute of Turkmenistan have led to a profound understanding of the geographical specificity of population formation in the settlement of Central Asia. Research work included the functioning of the settlement system on the territory of Turkmenistan; and for the long-term perspective, the author's participation in the compiling of the map of the USSR population (1976), in the national action plan for deserts in Turkmenistan, in the Regional program of scientific-technical progress on arid and semiarid zone, and participation in the preparation of the UN Convention concerning the struggle against desertification.

In Central Asia, settlement of the population has changed essentially. Here two periods can be singled out: before and after World War II. In the first period the urban population increased enormously; by 1940 it was 4.52 million having increased in comparison to 1913 by 135% and reached 26.5% of the total population. In the pre-war period many people from other regions, mostly the Russian Federation and Ukraine, came to Central Asia.

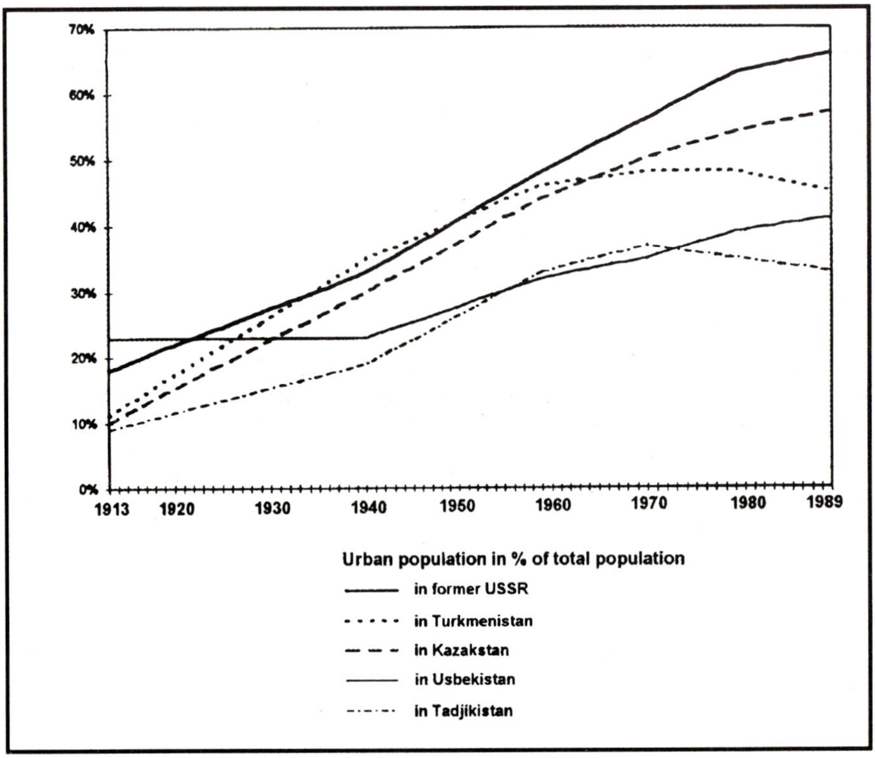

Fig. 4.1. The development of urban population in Central Asia

And among them there were qualified industrial and building workers, engineers and technicians, scientists and artists.

Demographic processes also influenced the settlement and dynamics of population in Central Asian countries. Except for Kazakstan, in the first period a high birthrate and a high natural growth rate of the native population were witnessed in Central Asian countries. Thus, from 1913 to 1940 the population in the Russian Federation increased by 22.4% and in Uzbekistan 51%.

W.W. II considerably influenced the population settlement. At the beginning of the war, many industrial and other enterprises were moved from the west to the east, especially to Kazakstan. The production capacities that were developed there spurred not only development of production forces, but a migration flow of population to Central Asia during the war and post-war years. In the post-war years, migrants resettled mostly to Kazakstan whose natural resources were intensively developed at that time, in particular fallow and unused lands.

The population in Central Asian countries has increased from 12.87 million. in 1913 to 93.9 million. in 1994 (Table 4.1. and Table 4.2.).

Table 4.1. Population of Central Asian states [in 1000]

	1913	1939	1950	1959	1970	1979	1989	1994
Uzbekistan	4334	6347	6264	8119	11799	15391	19905	22400
Kazakstan	5597	6082	6522	9295	13009	14684	16536	16800
Kyrgyzstan	864	1458	1716	2066	2934	3801	4290	4500
Tadjikistan	1034	1458	1509	1981	2900	3031	5109	5800
Turkmenistan	1042	1252	1197	1516	2159	2759	3534	4400
Total	12870	16624	17208	22977	32801	39666	49374	53900

The rural population has increased enormously, especially in the post-war years, because a high birthrate among the native population was maintained and because rural population was unwilling to move to towns. Thus, in Uzbekistan alone the total population increased from 5.4 million in 1959 to 12.3 million in 1991.

Many modern cities in Central Asia appeared in regions unsuitable for irrigated farming and even unfavorable for human living, but having advantages of location for the mining industry, nearby mineral deposits. Our brief overview of an uneven growth of population in Central Asian countries has revealed the following causes for this: different birthrates of the native population and different migration flows. These two factors varied from people to people in Central Asia, too. This led to a changed pattern of their settlement and to alterations of ethnic "mix" and historically established settlement forms (Fig. 4.1.).

4.1 Settlement

The territory of present-day Kazakstan had been occupied since ancient times by various nomadic tribes, and by the end of the XVth century they formed a single people - the Kazakhs. For many centuries the main occupation of Kazakhs was nomadic cattle breeding (rearing of Karakul and fat-rumped, wool sheep, horses, camels and large-horned cattle) which is of a self-sustaining nature. It determined the way of life of this people.

In the past, farming in Kazakstan was poorly developed. There were only small areas of irrigated ploughlands nearby wintering places, and more seldom rain-fed ploughlands were found on spring pastures. Kazakhs sowed mostly millet for their own consumption and alfalfa for horses. Only in the south of Kazakstan was small-scale commercial farming practiced; that was along the Tien-Shan piedmonts, in the Syrdarya, Talas and Chu river basins. The Uzbeks who had settled down in Southern Kazakstan long before created densely populated oases. Their main occupation was irrigated farming. Surplus farming produce they sold to their neighbors the Kazakh nomads.

Table 4.2. The population growth of Central Asian states [in %]

	1950-1959	1959-1970	1970-1979	1979-1989	1913-1994
Uzbekistan	29.6	45.3	30.4	29.3	417
Kazakstan	42.5	40.0	12.9	12.6	200
Kyrgyzstan	20.4	42.0	20.3	13.0	4210
Tadjikistan	31.3	46.4	31.1	68.5	460
Turkmenistan	9.1	13.3	8.1	28.1	322
Total	33.5	42.0	20.1	24.4	420

The southern margins of Kazakh settlement were the Karakum and Kyzylkum deserts. Using primitively natural pastures the Kazakhs moved throughout the year with their cattle from one pasture to the other. In the northern regions of what was then Turkmenistan, they populated together with the Karaklapks in the Amudarya river delta, and together with the Uzbeks on the right bank of the Amudarya.

The settled population was composed largely of the Tadjiks (descendants of the Khorezm people), and various ethnic groups speaking Turkish or Persian languages, as well as people from the nearby Pamir, as well as the already settled Uzbeks, Karakalpaks and some Kirghizs, Kazakhs and Turkmen.

The Tadjiks lived mainly in mountain valleys in the west of present-day Tadjikistan, on the left bank of the upper reaches of the Amudarya and Pyandj rivers, as well as on the slopes of the Ferghana and Zarafshan valleys, in many settlements in the Surkhandarya and Kashkadarya basins, in the Nuratin mountains and in towns.

For a long time the Uzbeks led a nomadic life making seasonal migrations from summertime to wintering places. Increase of loads on pastures because of the population growth and a growing number of cattle led to a more settled way of life and to a transition to farming of the former nomads, especially in such densely populated regions as the Ferghana, Zarafshan, Kashkadarya and Amudarya valleys. Gradually turning to a settled way of life the Uzbeks learned the art of farming and adjustment to a new way of life from the earlier settled population, largely the Tadjik people.

In the mid-19th century a large number of semi-nomads, the Karakalpak people living in the delta and on the right bank of the lower reaches of the Amudarya, started settling down. They were farmers and fishermen, but still they did not abandon cattle breeding completely.

The Turkmen people remained largely nomads and moved between the Caspian Sea and the Amudarya, the Mangyshlak and Northern Iran and Afghanistan. Each tribe had its own migration region. But among the Turkmen people there were fishermen who lived along the shores of the Caspian and the Amudarya, and also farmers who lived in oases of Kopet-Dag and in the Amudarya and Murghab valleys.

In the mid-19th century nearly all Kirghiz people were nomads. They populated mostly mountain slopes, inter-mountain depressions, plateaus of the Central Tien-Shan, the Eastern Pamir, and the upper reaches of the Chu and Talas, mountain margins of the Ferghana. They largely practiced mountain-pasture cattle breeding.

4.2 National Composition in Kazakstan

Central Asia is a very unusual region in its national composition, the everyday way of life, labor skills and traditions. A very variegated national composition of the population is typical of the Republic of Kazakstan. This diversity is a result of development of virgin lands suitable for farming by the Russian and Ukrainian people and other migrants as well as an inflow of international population to the constructed industrial centers and urban settlements. Already in the pre-war years in 1939 the population of the Russian people had doubled in comparison to 1926, and their share in the Kazakstan population was the largest (Table 4.3.). According to a population census of 1979 the Russians in Kazakstan were in the absolute majority in Karaganda, Eastern Kazakstan, and Northern-Kazakstan, Kustanai, Akmolinsk and some other regions, and in Almaty too. The Kazakh population was numerically dominant only in Adyraulsky, Kzyl-Orda, Ural and Shymkent regions.

In northern and eastern steppe regions of Kazakstan, apart from the Russians there were large groups of Ukranian people (Kustanai, Karaganda regions), and Germans (Karaganda, Akmolinsk and other regions). In some regions of Kazakstan, especially in the Karaganda Region many Tatars live. In the south groups of the Uzbek people have their homes (in the Chimkent region) and separate groups of the Uigurs. In the period 1959-1970 as a result of the natural growth of the Kazakh population and a decreased inflow of migrants of other nationalities, the share of the Kazakhs in the total population somewhat increased. From the 1970s when the inflow of people of other nationalities, and first of all the Russians continued diminishing a share of the Kazakh population has increased from year to year (from 1959 to 1989 by 9.5%), while the number of the Russians has been shrinking.

4.3 National Composition in Turkmenistan

In the national composition of Turkmenistan the share of Turkmens is 77%. Here the Uzbek (407,000 or 9.2% of the total population), Russian, Kazakh, Tatar and other peoples live.

Table 4.3. The dynamics of the national composition of population in Central Asia

		Population [1000]	Indigenous [%]	Russians [%]
Kazakstan	1926	6,503	57.1	19.7
	1939	6,094	38.2	40.3
	1959	9,295	30.0	42.7
	1970	13,009	32.6	42.4
	1979	14,684	36.0	40.8
	1989	16,536	39.5	36.2
Kyrgyzstan	1926	993	66.6	11.7
	1939	1,458	51.7	20.8
	1959	2,066	40.5	30.2
	1970	2,933	43.8	29.2
	1979	3,523	47.9	25.9
	1989	4,290	52.0	21.2
Tadjikistan	1926	827	74.6	0.7
	1939	1,485	59.6	9.1
	1959	1,981	53.1	13.3
	1970	2,900	56.2	11.9
	1979	3,801	58.8	10.4
	1989	5,109	62.0	7.7
Turkmenistan	1926	900	70.2	8.2
	1939	1,252	59.2	18.6
	1959	1,516	60.9	17.3
	1970	2,159	65.6	14.5
	1979	2,765	68.4	12.6
	1989	3,534	72.0	9.5
	1995	4,483	77.0	6.7
Uzbekistan	1926	4,446	74.2	25.4
	1939	6,271	64.4	11.5
	1959	8,119	62.1	13.5
	1970	11,800	65.5	12.5
	1979	15,389	68.7	10.8
	1989	19,905	71.4	8.3

The Russian population in Turkmenistan only inhabit the cities, especially in Ashgabat. In the period from 1926 to 1939 the Russian population in Turkmenistan increased three times, but after this its inflow became much less. As a result the percentage of the Russians in the population of Turkmenistan between 1959-1995 decreased by 10.6%, while the population of Turkmens constantly increases due to high, birth rates.

Turkmenistan is populated by many nationalities including some small, ethnic groups. Beginning from the late 19th century in towns and rural areas of Turkmenistan, small compact groups speaking Caucasian, Iranian and Turkish languages started appearing. They included the Persians, Afghans, Kurds, Baluchis, Turks and others. In search of money and security, separate groups of them arrived to southern regions of Turkmenistan. The most significant group among the minor nationalities are the Baluchis who live in the Mary velajat of Turkmenistan.

The so-called, small nationalities that can be found at present in all countries of Central Asia are mostly descendants of migrants. They are all bilingual or trilingual, they speak their native and Russian languages well. The most significant groups are Baluchis, Uigurs, Germans and others. As a result of their compact settlements, they preserve their family, marriage and other traditions, customs and ceremonies, language, and women wear national clothing.

Marriages are largely contracted within a group, although from time to time mixed marriages do happen. From the very beginning these groups have and still enjoy internal autonomy, and live in their special settlements. They have friendly relations with the Turkmen population.

The second significant group includes the population speaking Iranian languages (Parsuwan, Persian, and Zeban) that live compactly in the settlement Sarakhs, and a peasant community nearby Sarakhs. These groups (descendants from Iranian Seistan), like Baluchis, live very close and had practically no familial marital relations with the surrounding Turkmen population. They strictly preserve their language, clothing, customs and rites. They also listen to radio broadcasts and watch TV from Iran (Meshed). In recent years, however, young women and girls have started putting on Turkmen national clothing.

The Kurds live compactly in the village Bagir within the Ashgabat enclave. As a result of an intensive, natural assimilation, the Kurdish population is gradually decreasing (by 50% since the time of the 1926 census).

Until the 1960s small compact groups of Germans and Persians lived in rural areas in the Serakhs enclave of Turkmenistan. These peoples are not very numerous, and maintain friendly relations; they have common weddings, holidays, funerals and other national events, and they have no national conflicts.

Turkmenistan enclaves bordering on Uzbekistan and Kazakstan are populated by rather numerous Uzbeks, Kazakhs and other nationalities. In areas of their concentrated living (Kazakhs in the Turkmenbashi enclave of the Balkan velajate, Uzbeks in some enclaves of the Tashauz and Lebapsky velajates of Turkmenistan), they have schools and newspapers in their native languages.

4.4
National Composition in Uzbekistan

The national composition in Uzbekistan is very complicated, especially in the Tashkent Region where Russians, Kazakhs, Tatars, Koreans and other peoples are living. Large groups of Russians populate Ferghana, Samarkand and Bykhara regions, while in Tashkent they make a relative majority of the population. Many other ethnic groups (Tatars, Jews and others) like Russians concentrate largely in cities. Uzbeks prevail in nearly all rural areas, except for semideserts in the north where Kazakhs live and in southern regions where Tadjiks settled (Surkhandarya region, etc.). The greatest inflow of Russians to Uzbekistan as well as to Turkmenia was connected with the industrial development of the republic.

From 1926 to 1939 the Russian population in Uzbekistan doubled (from 241 to 727 thousands). In the post-war period this inflow decreased, and although the Russian population similar to other nationalities continued increasing in absolute figures, from 1959 on their share in the Uzbekistan population has been gradually shrinking as a result of high rates of natural growth of the Uzbeks.

4.5 National Composition in Tadjikistan

Tadjikistan is populated, apart from Tadjiks, by multiple groups of Uzbeks that live largely in the north-western and south-western regions of the country. Some regions near Parmir are populated by Kyrghyzs, while the Russian communities live mostly in cities making a majority in Dushanbe, for instance. In Tadjikistan the Russian population, like in all other countries of Central Asia, quickly grew till 1959. Later the growth slowed down and the share of Russians as well as other national groups (except for Uzbeks) shrank.

4.6 The National Composition in Kyrghyzstan

The dynamics of the growth of the national composition in Kyrghyzstan resemble in many respects Kazakstan. Here the Russian population has been growing. The Russians settled down in cities and piedmont regions for farming development. The western regions of the republic bordering on Uzbekistan (especially in the Osh region) are populated by Uzbeks, thus the population is constantly growing. There are also rather large groups of Germans, Ukrainians, Tatars, but their population in the recent decades has reduced, in general. As a result of such situations the share of the Kyrghyz population is growing.

4.7 The Development of Population

According to the 1926 census, the total population in Central Asia was 13.7 millions which was 0.8 million more than the population in 1913.

The demographic history of the peoples that populated arid regions is very complicated. In their historical development these ethnic groups had very different levels of socio-economic development.

Table 4.4. The birthrate, deathrate and natural growth of the population in Central Asian countries [per 1000]

	Year	Number of newborns	Number of deaths	Natural growth of population
Kazakstan	1985	25.1	8.0	17.1
	1990	21.7	7.7	14.0
	1991	21.0	7.9	13.1
	1992	19.9	8.1	11.8
Kyrgyzstan	1985	32.0	8.1	23.9
	1990	29.3	7.0	22.3
	1991	29.2	6.9	22.2
	1992	28.6	7.2	21.4
Tadjikistan	1985	40.0	7.0	33.0
	1990	38.8	6.2	32.6
	1991	37.7	6.0	33.7
	1992	32.2	6.6	25.6
Turkmenistan	1985	36.0	8.1	27.9
	1990	34.2	7.0	27.2
	1991	34.1	7.3	26.3
	1992	34.0	7.1	26.9
Uzbekistan	1985	37.4	7.2	30.2
	1990	33.7	6.1	27.6
	1991	34.6	6.2	28.4
	1992	33.1	6.3	26.6

Under different historical conditions the their fates were different e.g. there were periods in which cities flourished, while farming oases were sometimes destroyed after invasions.

The dynamics of the population in Central Asia became steady only in the 20th century. According to the 1939 census, the total population of Central Asia was over 16.6 million. In comparison to the pre-war year 1939, the total population in Central Asia in 1950 was only 584,000 more. This can be explained by the fact that during W.W. II many millions of people were killed, many died fighting at the front, and the birthrate everywhere was very low.

In the post-war years while life returned to its normal course and because of the reunion of families after men returned from the front, the birthrate rose slowly in comparison to war years. Thus, according to the 1959 census the population in Central Asia was 22.9 million or 38.2% more than in 1939.

In Central Asia in the period from 1939 to 1959 the population growth was very uneven. The population in Turkmenistan increased by 21.0%, while in Kazakstan and Uzbekistan the growth rate was above the Central Asian average. The Kazakh population increased by 52.8% which is due not only to a high growth rate of local population, but rather a high inflow of population, largely from the Russian Federation, in connection with intensified development of local, natural resources.

At present Central Asia is among those countries with high population growth rates. This is a result of the great diversity of natural, climatic, social and economic conditions, a multinational population, the unshakeability of everyday family traditions. The highest growth rates of population in Central Asia were observed between 1959-1970 when the growth rate was 42%, while during 1979-1989 this growth slowed down (24.4%).

In view of a complex of socio-economic factors, this unfavorable situation has become established in recent years in some Central Asian countries. Thus, in 1992 the natural birth rate per 1000 people decreased in comparison to 1985: in Tadjikistan from 33-25.6 people, in Kazakstan from 17.1-13.1, in Kyrghyzstan from 23.9-21.4 (Table 4.4.). At the same time in Turkmenistan and Uzbekistan a high birth rate was maintained (26.6-26.9 people per 1000 of population). The coefficient of infant mortality (up to 1 year of age) in Turkmenistan reached 45.9 per 1000 people.

The proportional rate of children and teenagers up to 16 years of age in the population of Central Asian countries is high. In this respect the most representative is the population in Uzbekistan and Turkmenistan where their percentage in the total population is 42.9% and 42.7%, respectively. The percentage of children and teenagers is especially high in rural areas in Tadjikistan, Uzbekistan and Turkmenistan and is 48.4, 48.3 and 47.1%, respectively.

Maintenance of traditions – many children in a family and a large percentage of population being young – facilitates the increase of a number of families. Here the stability of marital relations among native nationalities is also a factor which directly influences the number of children in a family. According to the 1989 census the number and size of families in Central Asia are as follows: the greatest number of families consisting of 7 members and more is in Uzbekistan (1,04 million families), then Kazakstan (340,000 families) and Tadjikistan (306,000 families). The average size of a family (family members living together) varies from 4 in Kazakstan to 6.1 in Tadjikistan.

The formation in Central Asia of newly independent states, a social crisis that affected the post-Soviet society, in particular inter-national conflicts, have essentially changed the factors and conditions of population migration on this territory.

4.8 Migration

For many decades Russia and Ukraine actively sent their people to the republics of Central Asia, especially in the period of active, industrial development. As a result, in Central Asia the population of the Russians and other peoples has grown enormously. But in the mid-70s the situation changed. Russia and other republics in the European part started attracting population from Central Asian republics. Thus, during a decade that passed between two censuses of the population (1979-1989) 1.81 million migrated back to Russia.

Migration losses from Tadjikistan in 1990 were 60,000, and during the 10 years between the 1979 and 1989 censuses 102,000. In the 1990s because of the intra-national conflict within Afghanistan the migration losses of Tadjikistan have been 20 to 26,000 yearly.

People migrating from Kyrgyzstan and Uzbekistan (50-100,000 a year) mostly go to the Krasnodar and Stavropol Territories, Rostov and Voronezh regions of Russia. Another large migration flow from Tadjikistan, Kyrgyzstan and Uzbekistan has migrated within Central Asia. The principal cause of migration was, until recently, instability in these countries

In Turkmenistan there has been no forced migration. They passed a law on double citizenship for the Russian population. This may be a reason why fewer Russians leave Turkmenistan and go to Russia. So in 1992 the migration flow was 13,188 people, then in 1993 only 9,312. At the same time a number of the Russians going to Turkmenistan from Russia has increased from 2,263 people in 1992 to 2,558 in 1993. This is indicative of stabilization not only of a migration, but the general, political situation in Turkmenistan.

There is an ample labor supply, a growth of labor resources and, as a result, more active competition for a workplace. Probably a greater part of the Russian population will leave Central Asia even if there are ideal inter-ethnic relations. Given a favorable scenario with conducive political and socio-economic conditions, the outflow of the population would be gradual and smooth, while under an unfavorable scenario the number of forced migrants would increase sizably.

4.9 Urban and Rural Population

According to the 1989 census the number of cities in Central Asian countries was 264 and urban settlements 458. The average share of the rural population, as of 1 January, 1991, was in Central Asia 46.3% in comparison to 43.7% in 1970. Today Kazakstan of all other Central Asian countries has the greatest urban population (56.7%). Until recently in Tadjikistan, Kyrgyzstan and Uzbekistan less than half of the population (31.4-40.3%) live in urban settlements. In Turkmenistan, the specific share of the urban population was 48% in 1979, and in 1996 it had become 45%.

Very important is the natural growth of the urban population proper, as well as the gradual transformation of large, rural settlements into urban settlements. In 1939 the urban population in Central Asia was already 26.5%. In Turkmenistan between 1939 to 1959 the share of the urban population grew enormously.

For the period from 1981 to 1991 36 new towns appeared in Central Asia, of which 22 are in Uzbekistan, 1 in Kazakstan and 5 in Turkmenistan.

Of the total number of urban settlements in Central Asia, the semiurban centers and small towns with settlements with a population up to 50,000 in which only 37.7% of the total urban population live, make the absolute majority (92%). More than 60% of the urban population are concentrated in large and major cities. The 1989 census revealed 2 cities with a million population: Tashkent and Almaty.

The population grows, first of all, in large multi-functional cities. And this process goes on most intensively in densely populated oases where development of agriculture and towns requires a large labor force, although settlement areas are also widening as a result of industrial development of mineral resources of deserts. Thus, between 1926-1991 the average population density in Central Asia had nearly doubled. If in 1926 population density was 5.9 persons per km^2, then in 1991 it was already 14.2.

General patterns of population settlement in Central Asia are very dissimilar, and the population distribution is spot-like. This reflects a clear-cut polarity in the general, regional population settlement. In suburban areas of large cities, in some rural areas of Central Asia the average population density is more than 600 people per km^2 (for instance, in the Khiva area of the Khorezm district in Uzbekistan), while vast expanses of deserts and mountains in Central Asia are either totally or hardly populated.

Towns and semiurban centers, workers' settlements in the desert zone are scattered and develop by themselves surrounded by vast territories of desert landscapes. At the same time there is a trend to setting up groups of settlements on transport crossroads and nearby water sources (water wells, pits, *sardabas*, etc.). The development of a water main network (e.g. Amudarya-Uchkuduk, Ashgabad-Erbent, Yaskhan-Nebitdag and others), and the water supply to desert pastures lead to the setting up of polycentric structures for compact living. This is conducive to the development of small oases for living in deep deserts.

At the same time it should be said that conditions and prospects of urban settlement development in a desert proper are more complicated and specific than in large oases which are a kind of "stronghold" for surrounding areas.

A mining industry is the main factor of development and growth of towns and settlements in desert regions. Settlements in deserts have remained small and scattered for long because of remoteness from oases and base centers and other difficulties. Purely urban working populations live in them, but these settlements are turning into towns more slowly than in other regions.

Urban settlements that appear in areas where mineral resources are mined or processed are scattered over a vast territory of deserts as a single oasis (Bekdash, Urkuduk, Beineu), or form whole "clusters" that are usually influenced by the earlier established base towns (Aktau, Nebitdag, Adyrau, Bukhara, Chardjou, Bairamaly and others). Under the industrial development of mineral resources, the conditions and prospects of population settlement vary perceptibly depending on the size of reserves of mineral resources, their location, scale and priority of development.

Some local groupings with a big, easily observable agglomeration have grown in Central Asia. Thus, for instance, the Tashkent capital region is a highly urbanized, integral territory actively influenced by Tashkent. Here the largest, monocentric agglomeration in the region with a total population of over three million has emerged. Adjoining Kelessy, the Saryagachsky regions of the Chimkent district in Kazakstan are also in the zone influenced by the Tashkent settlement system.

Ashgabat the capital of Turkmenistan, is in a different situation as to settlement and urbanization. Being one of the young capitals, Ashgabat up to the 1960s "attracted" all industrial settlements that appeared around it. Bezmein town that had sprang up not long ago now became its industrial satellite. In subsequent years Ashgabad continued drawing new enterprises and objects of social infrastructure.

Consequently, capital cities and district centers (velajates) play a leading role in the settlement system of Central Asia. They shape spatial systems of interrelated urban and rural settlements on the basis of production, and the cultural, labor and other ties. They become regional centers of groups of towns, urban and rural settlements with single transport systems, engineering communications, a single irrigation network, a system of cultural and everyday service of the population. At the same time some towns, for instance, Aktau, Nebitdag, Adyrau (former Guriev), Temirau and others are characterized by the absence of agricultural farming in the surrounding areas. They have a compact form of living in the midst of deserts.

4.10 Settlement Systems

Settlement systems in Central Asia formed a specific hierarchy of small towns and urban settlements that include: new, quickly growing towns and urban settlements that arose as basic centers of industrial development of mineral resources;

- urban settlements with a population of 5-10,000 that appeared on a free space in the zone of irrigation canals in the course of agricultural development of virgin lands
- small workers' and "service" settlements located in a desert on transport lines
- distant settlements in a desert for teams of shift workers

Therefore, settlement systems in Central Asia are formed differently and influenced by the following specific factors inherent only in arid areas of Central Asia:

- a sharp difference between natural economic conditions in a desert and oases, the complexity of historical, economic, national, ethnic and demographic processes, etc.
- lack of permanent surface waterstreams, a deficit of fresh groundwaters which impelled people to settle down within limited areas
- differences in levels of economic development of territories and urbanization among separate states and economic regions in Central Asia
- appearance of settlements in areas beneficial in economic and geographical terms and attractive for settling down, having water sources, mineral deposits, fertile soils, pastures and resources of building materials
- specific hierarchy and types of settlements in general ranging from the most minute nearby mines in a desert comprising one or several buildings and removable *yurtas* (with the population 3-5 persons), to cities with a population of a million
- uneven population of a territory: a sharp differentiation of the population density, a density of settlements and their populations between desert territories and oases with their rather compact settlement
- maintaining a high level of natural reproduction, especially of the rural population, and, in this context, rejuvenation of its structure
- a growing share of the urban population
- visible extension of settlement areas due to the development of natural resources, as well as the development of irrigated farming and distant range, animal husbandry
- a slow and infrequent transformation of urban settlements into towns

The specific share of rural population in Central Asia is constantly growing. In 1959 it was 38.4%, while in 1991 54.9%, and only in the last years has this unfavorable process nearly stabilized (the share of the rural population in 1989 was 53.0% and in 1991 54.8%).

The soaring growth of industry in Central Asian towns did not lead to diminishment of the rural population which continued increasing from 1940 to 1959 by 12.8%, and from 1970 to 1991 by 41.8%. In 1990 the specific share of the rural population in the overall population of Central Asia was 54.7%. Among the reasons responsible for such a situation the most significant were a considerable increase of areas where labor intensive crops were cultivated, and the development of new territories for irrigated farming. The population growth is also connected with significant, regional differences in the specialization of farms and the development level of productive forces in various regions of the arid zone.

4.11 Population Density

The highest density of the rural population is in oases and mountain valleys, and the lowest in deserts. An average population in rural settlements in Central Asia (in the Ferghana valley) reaches 1,200 persons, and an average spacing among them is only 2-4 km.

Apart from such densely populated regions in Central Asia there are also vast, very sparsely populated territories. For instance, in the Mangistau regions in Kazakstan an average population density is 2 persons per km^2, in the Adayrau region 3.9, and Dzhezkazgan region of Kazakstan 1.6 persons. In Turkmenistan, in general, this figure is 7.6 persons. Kazakstan is especially distinguished by its high urbanization and the industrial orientation of the economic development.

Beginning from the 1950s, extensions of areas with rural population became visible. Large canals built from the Amudarya and Syrdarya allowed an increase in irrigated areas by developing nearby territories (Table 4.5.), and created new administrative regions and rural settlements. In the last three decades new administrations units were formed in Uzbekistan (Navoiysky and Dzhizaksky); Karakumsky, Turkmenbashinsky, Gyaursky and Niyazovsky in Turkmenistan; and Yavansky, Kommunistichesky, Kabodiensky and Ilyichevsky in Tadjikistan.

The most important peculiarity in the geography of settlement in Central Asia is a high concentration of the population, especially rural, in small oases. Oasis-based settlement is most typical of Turkmenistan and Uzbekistan, i.e. the countries where deserts take up a considerable area.

Oases present a cultural landscape greatly transformed by human activity. This is a territory with a very high concentration of economic activity and population. Historically, oases appeared in places where rivers came out from mountain ridges or in lower reaches of rivers where there were possibilities for practicing irrigated farming. Oases are a purely geographical phenomenon: against an arid background they are, in general, areas of intensive life. A river network in these regions determines the main outlines and a pattern of location of production forces, a framework of a regional settlement of population.

Mostly rural or rural-urban populations formed oases where large villages are integrated into and intertwine with small towns of "the rural origin", and where there is not a clear-cut border between a town and "non-town". Such a type of settlement is an undivided unity of a town and rural settlements.

An average spacing between urban settlements in some oases is less than 20 km, and between rural ones at least 5 km. Such conditions are conducive to the formation of integrated, rural-urban systems of settlement. In this context it can be said that each oasis is a single, super-settlement, a kind of a rural-urban agglomeration, while several interlinked oases form a kind of megalopolis.

Table 4.5. Development of the urban and rural population (1959 and 1989 censuses)

	Percentage of urban population		Percentage of rural population	
	1959	1989	1959	1989
Kyrgyzstan	34%	38%	66%	62%
Turkmenistan	46%	45%	54%	55%
Uzbekistan	34%	41%	66%	59%
Kazakstan	44%	57%	56%	43%

At the same time mountain and desert regions are characterized by a sparse network of small settlements, a spot-like territorial organization of production and population. Narrow, colonial settlement of the population is typical for these regions. Numerous towns and settlements in these regions are beyond the main, generally populated lines and areas.

In general, a macro-regional settlement structure in Central Asia includes oasis, mountain valley and sparsely populated mountain and desert regions. A territorial combination and interaction of the latter determine the geographical originality of the region. In a wider sense, the territorial structure of the economy and settlement in Central Asia is formed by densely populated oases and mountain valley regions, on the one hand, and sparsely populated desert and mountain regions, on the other. Correlation of these two superstructures gives a geographical factor of region development. Against such a background the role of a particular piedmont zone is drastically enhanced historically and geographically. It is through this zone that a linear component of a supporting framework of settlement runs, and it also comprises main, nodal elements of this framework in the form of densely populated oases and rural-urban agglomerations.

5. Desertification, Its Consequences and Control Strategy

A.G. Babaev, I.S. Zonn

ABSTRACT: The causes of the modern occurrence of desertification are defined and explained. Anthropogenic factors such as wasteful farming, uncontrolled industrial development, and tourism have played a critical role. More differentiated, regional research linked in turn by international coordination is required to identify exact signs of this process. The UN has played an especially important role in highlighting the urgency of the challenge, and the Convention to Combat Desertification is foregrounded, along with The Desert Institute's special role in the global campaign. Solutions can only be found within a socio-economic context which should not be separated from the ecological dimension.

Desertification is one of the most dangerous ecological and socio-economic issues that faces the world community on the threshold of the 21st century. It is quite understandable that it causes mounting alarm among scientists, business circles and governments of many countries, and this concern is indicative not only of anxiety amongst the world public given the genuinely critical state of the natural environment, but also an endeavor to render assistance to the populations of Africa, Asia and Latin America who have suffered from desertification.

Desertification processes affect largely arid territories that take up about 30% of the land area on the globe on which there is about 145 million ha of irrigated lands, 170 million ha of rainfed lands and 3.6 billion ha of natural pastures.

In the last 25-30 years arid lands have been subject to a special anthropogenic load, when in some countries they started wide-scale development in their efforts to find new sources of mineral resources, land areas and biological resources. Extensions of irrigated lands, increases of cattle population, intensive procurement of natural forage and wood for fuel, the introduction of agro-industrial methods of land development were the reasons for a drastic disturbance of an ecological-resource balance of arid landscapes.

It is generally known that the nature of arid lands is very fragile and vulnerable, that is why any unwise human actions in the development of their resources leads to serious ecological stresses that are called "desertification". Desertification is a product of a complex interaction between a socio-economic system (diseases, poverty, hunger, unstable economics) and natural factors (drought, water erosion, soil salinity, degradation of vegetation). These two domains can easily intensify each other, but they also have a feedback in the course of desertification.

As is already known, the present-day arid zone is both a sphere of vast potentialities for development but also a place where intensely contradictory social, ecological, and political factors clash.

International publications provide different interpretations of the term "desertification". After generalizations and analysis of all proposals, the following definition of desertification is given in the text of the International Convention to Combat Desertification: *"desertification is the degradation of land in arid, semi-arid and sub-humid areas caused by various factors, including climatic changes and human activities."*

Arid territories possess significant natural and labor resources. The study and gradual development of these riches can be considered as a major strategic reserve for the progress of the human species. At the same time it is in the arid regions that people live in awful poverty and suffer from hunger and diseases. In such conditions provision of any possible help to countries and peoples that happen to be in the zone of intensive desertification is a moral duty of the world community and, above all, of scientists who have amassed significant experience of how to combat this dangerous phenomenon.

Desertification can be observed in any climatic conditions, and the rate of its development depends on natural regimes there. In arid regions with a dry and hot climate, the desertification processes accelerate and can reach catastrophic proportions, while in humid, hot and moderately warm regimes they go on much slower. Recurring, severe droughts can induce an acceleration.

Desertification is usually represented as degradation of land, water, vegetation and other natural resources in conditions of demographic and ecological stress. This is not a frontal process of natural encroachment of a desert on marginal zones, but a sporadic and gradual spreading of degradation processes to different parts of the adjoining humid zone (steppes or savannas). Such desertification is caused mostly by uncontrolled human activity which does not take into consideration diverse and fragile interactions of finely balanced, natural components that form the biological productivity of arid territories (relief, soils, flora and fauna), and that are very sensitive to loads produced by land use. It is revealed in changed compositions of vegetation cover and animal populations with attendant decrease of their biological productivity, the worsening of a water regime, the development of devastating erosion and deflation processes.

5.1 The Desertification Problem

Desertification is not something new for people. It has been observed since ancient times over the whole world, and was historically related to unwise land use.

The development of land use in arid regions abounding in victories, discoveries and dramatic errors played a certain role in the history of many human civilizations. Middle Asia, African coast of the Mediterranean, Central Asia: this is a far from complete list of major regions that suffered from desertification in the past.

Desertification development was spurred not only by uncontrolled, large-scale withdrawal of resources of arid lands. People, on the one hand, are creators of desertification, and its victim, on the other. Formerly there was a widespread belief that desert resources are inexhaustible, and they did not often properly appraise their actions which could lead to extensions of desert boundaries and new losses of lands.

The clearest example of quick desertification is the Aral Sea basin which became a victim of the big, water development projects that had promised so much.

It was found that of 45 specified causes of desertification the unwise management of water, land, vegetation, mineral resources make up 39 and only 6 are due to natural processes. Desertification often starts during severe droughts when land, subject to an increased load from land users, cannot properly resist human pressures.

The desertification issue becomes still more complicated in the face of optimistic evaluations in wet years, which then cause an excessive load on lands in subsequent unfavorable years, in turn invariably leading to a more severe desertification. Besides, technological novelties transferred here from humid regions without proper account of the necessity of maintaining specific features of ecological equilibrium in arid ecosystems often result in intensification of desertification processes in areas of their application. Therefore, the main quantitative factor spurring desertification is population growth, and the qualitative factor is the application of new and ecologically, unsatisfactory practices of the exploitation of natural resources. These anthropogenic factors are closely interconnected with other social events, such as break-up of widely applicable, traditional practices of land cultivation, the inadequacy of social structures, etc.

The desertification problem acquired a worldwide significance in the period of tragic events of 1968-1973 when a devastating drought that was later called "the great" affected the Sahel zone in Africa to the south of the Sahara. Its consequences for such countries in this zone as Senegal, Mali, Upper Volta, Niger, Chad, Mauritania and others were unquestionably dramatic.

Geographical differentiation from region to region of the scale and nature of human desertification in developing countries is dependent, first of all, on the economic activities of local populations which apply, on the one hand, traditionally established practices, and, on the other, modern technologies that are often unsuitable for specific conditions of a fragile, arid ecosystem.

Some practical activities can cause particularly negative consequences in arid conditions. First of all, this is the degradation of vegetation cover and attending soil degradation (erosion and deflation) arising due to overgrazing of cattle, and cutting of plants for fuel by the local population. It is well-known that pasture cattle breeding is the principal, land use in deserts. Earlier the most intensive, human impact of a local nature on the natural environment was only on nearby drinking points; however, this started widening rather quickly with the growth overall of cattle population, and so acquired not merely a local dimension but also a regional scale. The migratory nature of such cattle breeding increases the load on arid pastures, and makes it more difficult to introduce rational methods of its organization. Vegetation around settlements is actively cut by the settled cattle-breeders and farmers for fuel, thus allowing desertification to creep to margins of their settlements.

5.2 Socio-Economic Aspects

Prevention of desertification and its control includes in many cases a complex of mutually interrelated social, economic and engineering actions that are incorporated into a single plan of socio-economic development of some natural-territorial complex.

This deserves special mention because quite often during practical activities people pay attention only to a technical aspect of this problem overlooking here its ecological and socio-economic aspects.

5.3 Irrigation Farming

As is known, irrigation provides the most productive basis for agriculture in arid regions. It ensures higher stability of farming practices, alleviates the danger of droughts and uncertainty about the future. At the same time, irrigated farming is the most complicated agrosystem controlled by humankind, and its mismanagement often results in such dangerous processes as secondary salinisation, alkalisation and waterlogging of irrigated and nearby lands. At present the area of saline lands in the world is about 900 million ha. Special anxiety is stirred by progressing salinization in developing countries in the arid zone where economic factors prevent the rational construction and operation of land reclamation systems using informed modern engineering. Among processes of desertification caused by people, deflation of sands fixed with vegetation that develops with overgrazing of pastures, cutting and rooting out of vegetation for fuel deserve special mention.

5.4 Industry

Industrial forms of development of natural resources produce negative impacts on arid ecosystems. Industrial development leads not only to the setting up of "technogenic zones", deterioration of geochemical characteristics of a landscape, the appearance of "lunar landscapes", which spur the processes of soil erosion and deflation, diminishment of its fertility, but is connected too with the building of new settlements, the construction of roads, irrigation structures, gas- and oil fields. As a result, great quantities of machinery is transported to deserts. Apart from the unquestionable economic boost to desert development, these processes bring a number of negative impacts.

5.5 Tourism

Uncontrolled tourism, recreational activities of the native population (especially urban population) connected with uncontrolled picking of wild flowers and other plants, and disturbance of animals in their critical periods are also factors that are conducive to desertification, because they lead to the diminishment and even complete destruction of rare and economically valuable populations of plants and animals.

5.6 Military

In the recent years military actions, tests of nuclear and strategic weaponry, maneuvers of ground troops have become a powerful factor in desertification in marginal regions of the world.

5.7 Control Strategy

A continuous growth of the population in arid regions, the further development of industry and agriculture will, of course, involve intensive use of natural resources of arid lands in the economies of many countries. The main accelerator here will be energy and foodstuff resources. Their use will help develop contiguous branches, such as transport and municipal construction, irrigated farming, animal husbandry and others. It should be remembered here that the present development phase demands strict rationing of their planning. Exceeding admissible loads on their ecosystems in arid regions is an active desertification progress.

The goal for future researchers is to undertake zoning of desertification types in order to outline territories, on the one hand, for application of certain technologies to control desertification, and on the other, for prediction of likely intensification of this process.

Very important for preservation of a fragile desert ecosystem will be scientifically valid development and planned implementation of integrated global, regional and national programs aimed at desertification control.

Despite the fact that we have already learned a lot about methods of using these lands and technologies developed for this purpose, we still do not have a very clear understanding of how this knowledge can be used with regard to existing social, political and economic constraints. While developing these programs it is necessary to undertake an all-round appraisal of economic and social losses from desertification so as to reserve investments for their implementation. But one thing is clear: that this is dependent, to a great extent, on the execution of scientifically based and technically feasible, regional and national projects.

By now the foundations of scientific knowledge on world deserts have been created, and great experience in desertification control has been amassed.

5.8
Rational Use of Natural Resources

Of course, development and advances in science and technology give rise to a certain optimism. New scientific and technical possibilities to fathom the mysteries of nature, and the more rational use of its resources have been revealed. But at the same time many questions that are difficult to answer arise. And the first question concerns the economic advisability of each action. That is why it is very important in the face of present-day demands to be able to compare approaches to desert development and desertification control using many parameters, and, first of all, to compare capital costs with awaited benefits. A very important task that can be resolved in conditions of present-day progress in science and technology is to find not only sound, but relatively cheap methods of the use of biological and mineral resources of a desert without disturbance of its ecological equilibrium.

Such analysis will help establish the ultimate possibilities of further economic and social development of people in arid regions without any costly projects. In this context a research program on the development of scientific foundations for the rational use of natural resources of deserts and control of desertification is absolutely vital, and scientists and practical workers must rally their forces to implement it.

5.9 International Cooperation

The global issue of desertification control was first discussed at the UN Conference that was held in Nairobi in 1977. This Conference made retrospective assessments and adopted the 'Plan of Action to Combat Desertification'. From the very beginning this problem was perceived by the scientific community as one of the most critical issues involving the interests of countries and peoples not only in the arid zone, but in neighboring territories. Seeking its practical solution they set themselves a task to transfer their experience of research and development of arid land resources to countries affected by desertification. For this purpose, the period from 1979 to 1992 the Institute of Deserts (Ashgabat) organized international, scientific training courses on desertification control within the UNEP framework.

The wide use of the natural and economic potential of arid territories for the development of countries and peoples is possible only in conditions, firstly, of integrated researches and, secondly, an integrated approach to the development of desert resources only on the basis and within the framework of general, socio-economic progress. Here it is very important to take everything rational both from traditional development methods and from modern technology, while curbing as far as possible the latter's negative consequences for the natural environment. The leading strategic principle in desertification control is paradoxically not keeping the arid regions intact, but rather fostering their development with a view to improving the living standard of the local population on the basis of rational, eco-management.

Wide international cooperation in desertification control which started in the late 1960s not only significantly extended our knowledge of the processes of natural, environmental degradation in arid conditions, but also allowed an in-depth analysis and appraisal of the methodological approaches applied in the world to study global and regional specific manifestations of desertification processes, and to share positive experiences in desertification control with the countries that need scientific-technical support and help.

In this context the documents adopted in 1992 by the UN Conference on Environment and Development (UNCED) in Rio de Janeiro are of special importance. Apart from other vital, nature conservation problems the Conference adopted the proposals of developing countries on preparation of the International Convention to Combat Desertification (CCD). The conference considered this Convention "as a new important instrument to mobilize political, human, scientific and technical resources to combat desertification and for alleviation of the consequences of drought."

5.10 The Convention to Combat Desertification

In December 1992 the UN General Assembly established the Intergovernmental Committee to prepare the International Convention to Combat Desertification with headquarters in Geneva. Within rather a short time period, the Intergovernmental Committee succeeded in drawing attention to and directing efforts of many international, regional and national organizations, establishments, scientists and experts to the preparation of the Draft International Convention to Combat Desertification. As a result the final version of this most important critical document was prepared. This Convention was signed in Paris on 14-15 October, 1994. It was signed by more than 100 world countries, and they expressed their resolve to do everything to attain the goals of this Convention in the interests of present and future generations.

This Convention pursues the main objective of "Desertification control and alleviation of the consequences of drought in the countries that suffer most from a severe drought and/or desertification, especially in Africa, by taking efficient measures at all levels and signing agreements on international cooperation and partnership within the framework of an integrated approach which is in conformity with the relevant Agenda for the 21st century, and directed to securing steady development in affected regions."

This Convention also incorporates a strategy for successful survival in complicated arid conditions. Institutional and state bodies, as well as individuals must see in this Convention the guiding principles of a long-term policy to combat desertification caused by the human species. And if this policy finds its realization in joint practical efforts, it will help in consolidating national, regional and global actions that have been already taken within the UN frame and beyond it.

The main goal of this Convention is to prevent and curb a process of desertification, and where possible to restore biological productivity of degraded lands within ecological limits to improve the living standard of the local population.

Central Asian countries, such as Kazakstan, Turkmenistan and Uzbekistan, the territories of which are largely in the arid zone, joined the Convention and started preparation of their national actions programs for coping with this global problem.

5.11 Integrated Research

The experiences of the Desert Institute accumulated during many years was used in the course of preparation of the Regional and National Programs for Combating Desertification. The close attention of the UN and the world community to the desertification issue is one more confirmation of the fact that it has assumed a global scale, and threatens many countries in the arid zone.

The analysis of levels and rates of desertification that has been carried out in recent years revealed insufficiency of basic knowledge on monitoring of desertification processes, and their interaction with climate variations, periodic droughts and anthropogenic factors.

The results of integrated researches and appraisal of the processes governing the development of natural events in the arid zones become, in fact, a basis for taking optimal decisions in the context of this problem. Proceeding from joint studies of natural and socio-economic phenomena, it became possible to build a complete model of the present state of ecosystems in arid zones, and to resolve more correctly the problems of forecasts connected with the further development of arid lands and the combating of desertification.

Global desertification processes will be understood completely only if regionally differentiated, geographical phenomena are identified.

As the danger of desertification arises as a result of joint action and development of natural and socio-economic factors, it is necessary to consider them as a single phenomenon and study in more depth their action, interrelation, interdependence and contradictions. As the processes underway in the biosphere and geosphere are of global significance, their resolution depends on effective international cooperation.

Proceeding from the results of natural and socio-economic investigations, it is possible to build models of the interaction nature-society. This will allow the development of design proposals and the forecasting of solutions for the regulation and control of these interactions. In these terms it is very important to study the consequences of long-term impacts on the natural environment of land reclamation: irrigation, grazing of cattle, cutting firewood, engineering measures, etc. All the above requires a geographical forecast i.e. an ecological and social expertise.

5.12 Outlook

In the future very serious, quantitative and qualitative changes will be undoubtedly observed in the interrelations of nature and society, and in the ecosystem of the arid zones.

Analysis of the efforts of international and national organizations to implement the Plan of Action for Combating Desertification has indicated that actions recommended by the Conference are inadequate. Many developing countries face a progressing degradation of the natural environment in combination with traditional loads on their resources. In many developing countries a degradation of the natural environment is associated with high rates of industrialization and urbanization combined to some extent with negative economic affects and poverty.

Irrigation in itself is costly, although water is supplied to farms free of charge. If one does not set a price on water and land, and does not learn how to save fresh water, then spills and enormous losses of water to seepage and evaporation will be observed everywhere.

There is always a certain social dimension to project planning. A project is characterized by its considerable scale, its influence on the local economics and ecology. But unfortunately, administrative departments take purely technical and narrowly departmental approaches to these projects, and do not always take into consideration the consequences of its implementation. At the same time the slightest errors in implementation of these projects lead sometimes to serious negative consequences that outweigh awaited benefits.

Due to anthropogenic pressure the natural environment is strained to its maximum. And this poses the question as to whether risk-taking is justified knowing that in the near future we will step over an ecological limit.

For a better understanding of the complicated mechanism of desertification processes and the development of an ecological policy, it is necessary to set up an information support center and systems observations coordinated at various levels and using results of scientific investigations. During the 35 years of its activity the Desert Institute has accumulated unique experience. It was established in order to concentrate scientists and resources for the development of scientific and practical foundations of integrated study and rational management of resources of arid lands and the control of desertification processes in Central Asia. In this case, science itself become the most important component in the search for a way to combat desertification, and to ensure steady development in conditions of vulnerable landscapes. The conditions for the successful implementation of the International Convention on Desertification Control and alleviation of the consequences of drought remain a profound challenge.

6. The Monitoring and Forecast of Desertification Processes

A.M. Babaev, N.G. Kharin

ABSTRACT: Various monitoring methods for desertification are analyzed and assessed. Various parameters are explained for a desertification risk formula and a vegetation index. The Central Asian republics found it necessary to set up complex data banks using the latest technologies. Space photographs are the most useful tool for studying global and regional changes. The construction of the survey and the parameters included are explicated in detail. Forecasting is inextricably linked with surveying, and the principles on which predictions are based are explained. It appears that desertification processes will intensify in the near future.

Researches conducted by the Desert Institute of Turkmenistan have revealed that 66.5% of the territory of Turkmenistan, 60% of Kazakstan and 59% of Uzbekistan are affected to various degrees by desertification processes. In such situations it is necessary to take actions to restore and improve the natural environment. And scientific researches are very important here. The priority task in developing measures to combat desertification and indeed implementing them is the appraisal and mapping of desertification processes. The Deserts Institute using recommendations of FAO and UNEP (Kharin 1983) developed just such a method for appraisal and mapping of desertification processes applying remote sensing. On the basis of this method they prepared desertification maps of the Aral sea basin, Mongolia, Mali and Afghanistan (Figs. 6.1.-6.3.). Special criteria were developed for appraisal of each type of desertification processes (Tables 6.1.-6.5.). These tables also give sources that were used in development of some criteria. The following abbreviations were used:

- AP aerial photographs
- AD analytical data
- SP space photographs
- FI field investigations
- TOP topographic maps
- TM thematic maps
- LIT specialized literature
- ST design and statistical data

Table 6.1. Criteria of vegetative cover degradation evaluation

Aspects	Criteria	Desertification status				Source
		Slight	moderate	intensive	severe	
PS	State of vegetative cover	Climax seigatey changed communities	long existing secondary communities	short-living secondary communities	almost default of vegetative cover	FI, LIT, SP, AP
DR	Contemporary productivity (% of potential productivity)	> 90	80-90	30-60	< 30	LIT, ST
	Decrease in biomass	< 10%	10-25%	25-50%	> 50%	LIT, FI
	Annual increase in pastures' degradation	< 2,5%	2,5-5%	5-7,7%	> 7,5%	SP, AP, TM
	Annual forest felling (% forest areas)	< 2,5	2,5-5	5-7,5	> 7,5	ST, LIT
	Annual decrease in pasture forage	< 1%	1-4%	4-7%	> 7%	LIT, ST
ROD	Ecosystems' stability	Stable ecosystems of clay, loamy and gravel desert	Relatively stable ecosystems of forested deserts, and ecosystems developed on sandy soils and gentle mountain slopes	Unstable ecosystems of mountain slopes of average steepness, ecosystems of foothill loamy plains	Unstable ecosystems of sand desert, ecosystems of steep stone slopes	LIT, ST, TOP, SP
	Potential possibilities for new lands cultivation (% yearly per project)	< 3	3-5	5-10	> 10	ST

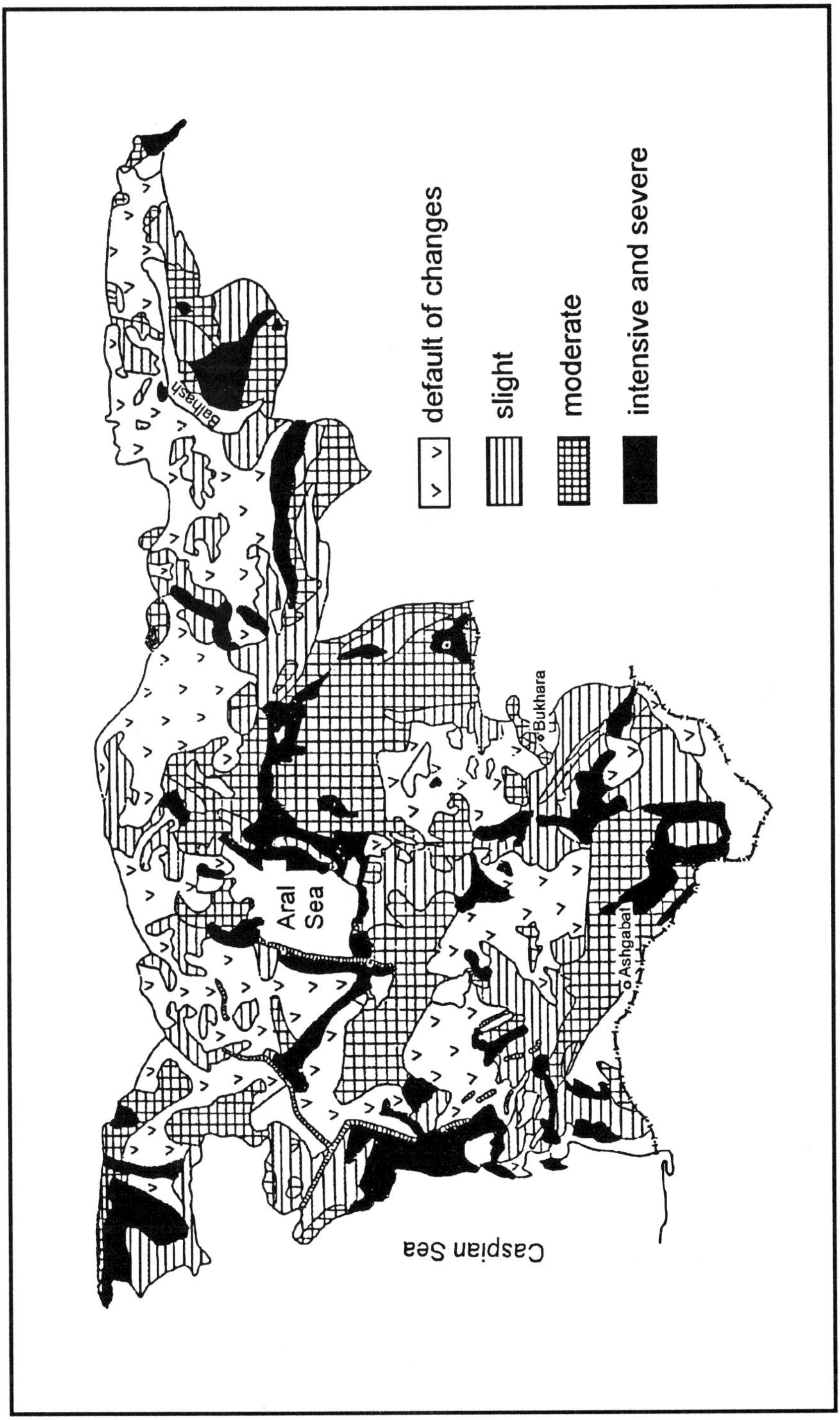

Fig. 6.1. Map of current desertification in Central Asian arid zones

Depending on coverage of a territory they can be local, regional, zonal and universal. So, with the increase of a mapping area the accuracy of criteria decreases. For preparation of the world desertification map it is necessary to apply universal criteria of desertification. Unfortunately, they have not as yet been developed.

6.1 Desert Risk Formula

Desertification is characterized by a set of parameters. Maps show types of processes (degradation of a vegetation cover, deflation of sands, water erosion, technogenic factors, salinization of irrigated lands and others), and such aspects of desertification as a background level, present state (PS), desertification rate (DR), internal risk of desertification (ROD), influence produced by animals on the natural environment (IA) and degree of anthropogenic impact (AI). The total risk of desertification (TRD) can be determined by the following formula,

$$TRD = PS + DR + ROD + AI$$

Maps of total risk of desertification show these criteria with regard to intensity of processes (classes of desertification). Four classes of desertification are usually applied: weak, fair, strong, very strong. On the maps the desertification types are denoted by non-scale signs and the total risk of desertification by different colors: a weak desertification class – yellow, fair – orange, strong – brown, very strong – red. Content of each contour is denoted by letters that include a point system of appraisals. Therefore, a map of total risk of desertification includes five separate maps.

6.2 Green Index

A vegetation index (a green index) can be used as an additional parameter in appraisals of desertification processes. This index was used by the Institute of Deserts while preparing desertification maps of Mali. Degradation of tree vegetation in savannah was appraised by the vegetation index and productivity. A vegetation index was estimated by space photographs from Meteor-Priroda (MSU-SK). It is proposed to make desertification maps of large, medium and small scales. In recent years the desertification processes in the Central Asian region have intensified.

Table 6.2. Criteria of deflation evaluation

Aspects	Criteria	Desertification status				Source
		slight	Moderate	intensive	severe	
PS	Surface state (relief)	Deflation of precipice erosion and hollows up to 25% on a outline area	of 25-50% precipice hollows area, signs of ripples on ungrown surface	Mass formation of falling slopes on mobile relief forms	Barkhan relief, default of vegetation and turf, carring out of particles from the whole surface	FI, AP
	turf area	30-50%	10-30%	< 10%	default of turf	FI, AP
	area covered by bushes	20-50%	5-20%	1-5%	< 1%	FI
	area covered by grasses	40-80%	10-40%	5-10%	< 5%	FI
	The depth of flowing out of unsandy soil in % out of a rooted layer capacity	< 10	10-25	25-50	50	FI
DR	annual increase of eroded area	< 1%	1-2%	2-5%	> 5%	TOP, SP, AP
	Annual blowing out of soil [1000 ha]	< 0,5	0,5-0,1	1,0-5,0	> 5	LIT
ROD	Soil group on mechanical composition	Clay, heavy loam	Average and light loam	Supesand	Sand	TM, LIT, AD
	Anthropogenic effect on sandy surface	Breakage of relief streamlines	Cutting off of bushes and overgrazing	Destroy of turf	Land works without sand anchored measures	SP, AP, FI

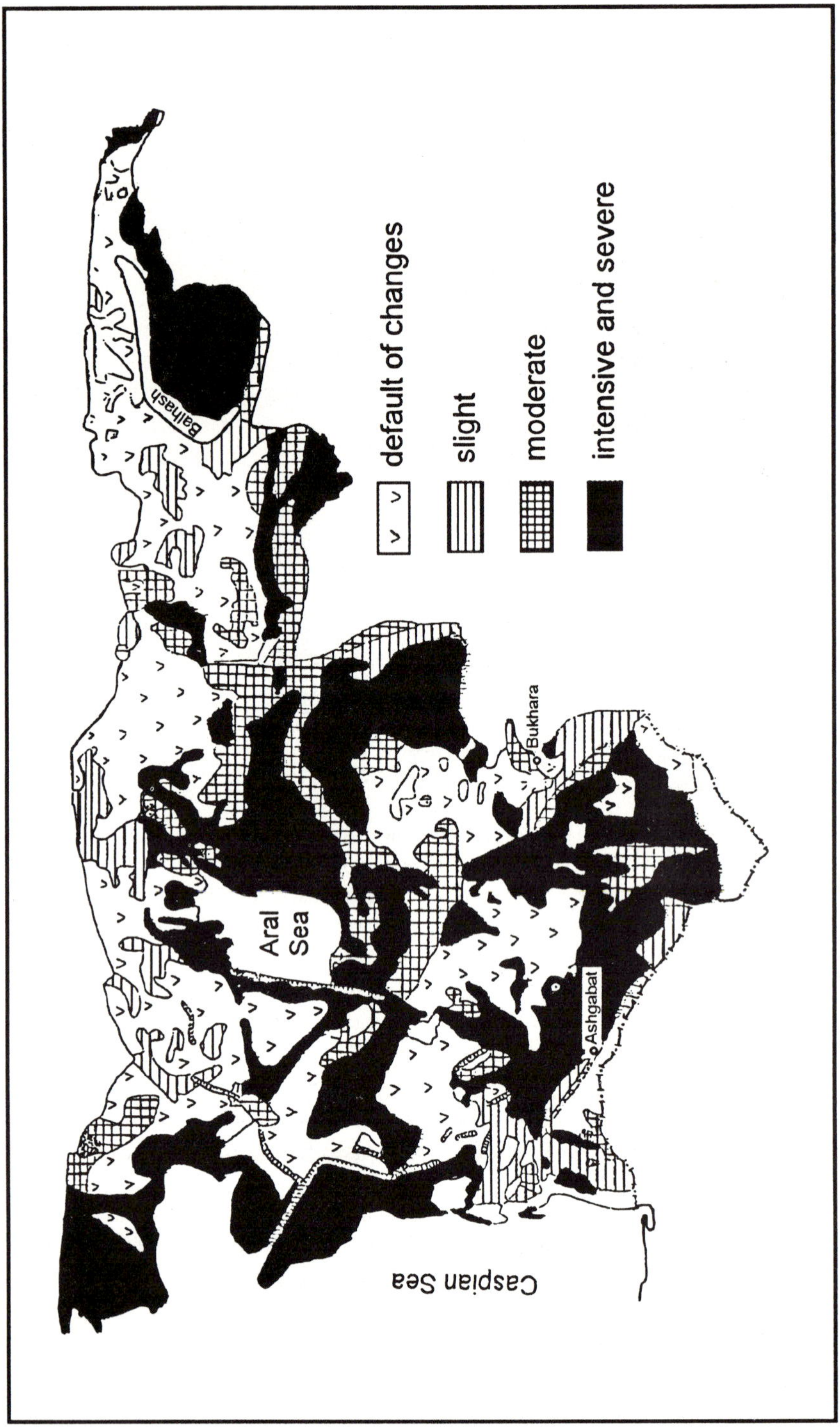

Fig. 6.2. Internal desertification danger of Central Asian arid zones

6.3 Monitoring Network System

Quickly changing desertification processes make it necessary to organize a special monitoring system. The obtained information will help appraise the state of the national environment in the region quickly and objectively. Development and implementation of desertification monitoring will essentially extend data files that should be transferred, processed, stored and used while coping with the desertification control problem. And it is general knowledge that this cannot be done without an appropriate information and monitoring network based on modern, computer information technology.

Scientific institutions in Central Asian countries created banks of data on the present-day state of the natural environment in order to supply relevant information to scientists.

The set goals can be attained by:

- application of high-capacity computers for collective use with developed periphery, mini and macro PCs, communication means and data transfer
- wide application of novel technologies for information interpretation, software for input, storage and transfer of information
- ensuring wide interaction of information systems of various levels cooperating in information identification, exchange of information, and its transfer via communication channels.

The introduction of an information and monitoring network to investigate desertification processes is possible if centralized, automated information systems of these countries are used. Connection of all systems will facilitate prompt exchange of necessary ecological information via telecommunication channels among all countries.

It is known that it is impossible to resolve a problem of rational, natural management without concrete and comprehensive information. To cope with this problem it is necessary to organize an integrated monitoring system of desertification processes. Monitoring is understood here as a system of observation, control and supervision of the natural environmental that is accomplished in various scales. This term also includes reliable forecasts of the development of geosystems as one cannot control processes without such predictions.

For the study of desertification processes in Central Asia the most suitable are regional and local levels of monitoring that provide objective and diverse information on particular territories. Regional monitoring allows specification of the lifetime of various desertification processes, appraisal of their spatial-temporal dynamics and validation of their predicted development.

Table 6.3. Criteria of water erosion evaluation

Aspects	Criteria	Desertification status				Source
		slight	moderate	intensive	severe	
PS	Surface characteristics	Gravel and stones cover is less than 10%	Stones and boulders cover is 10-25%	Boulders and rocks cover is 25-30%	Boulders, rocks and exit of main rocks cover is > 50%	FI, TOP
	Types of water erosion	Flat flowing and stream erosion from slight to moderate	Flat flowing and stream erosion from moderate to severe	Severe erosion: Flat flowing and ravine erosion	Severe erosion: Flat flowing and severe erosion	SP, AP
	Number of water pits, *grooves* and *ravines* per km	Water pits and grooves are of less than 5	Water pits and grooves are of 6-10	*No number*	Water pits and ravines are of more than 10	AP, SP
	% of bare soil	< 10	10-25	25-50	> 50	AP, SP
	Capacity of soil layer	90 cm	90-50 cm	50-10 cm	10 cm	FI
DR	Annual increase of eroded area	< 1%	1-2%	2-5%	> 5%	FI, LIT
	Soil loss, t/ha yearly	< 0,5	0,5-10	1,0-5,0	> 5,0	FI, LIT
ROD	Relief	slightly waved	hilly	mountainous with gentle and steep slopes	mountainous with steep slopes and fallings	TOP, AP
	Steepness of slopes	< 5°	5-15°	15-30°	> 30°	TOP, AP
	Area covered with trees and bushes	> 30%	30-15%	15-5%	< 5%	AP, FI
	Area with developed turf (for zones of grassy associations prevalence)	> 50%	50-30%	30-10%	< 10%	AP, FI
	length of rills and gullies	< 0,5 km/km²		> 0,5km/km²		TOP, AP

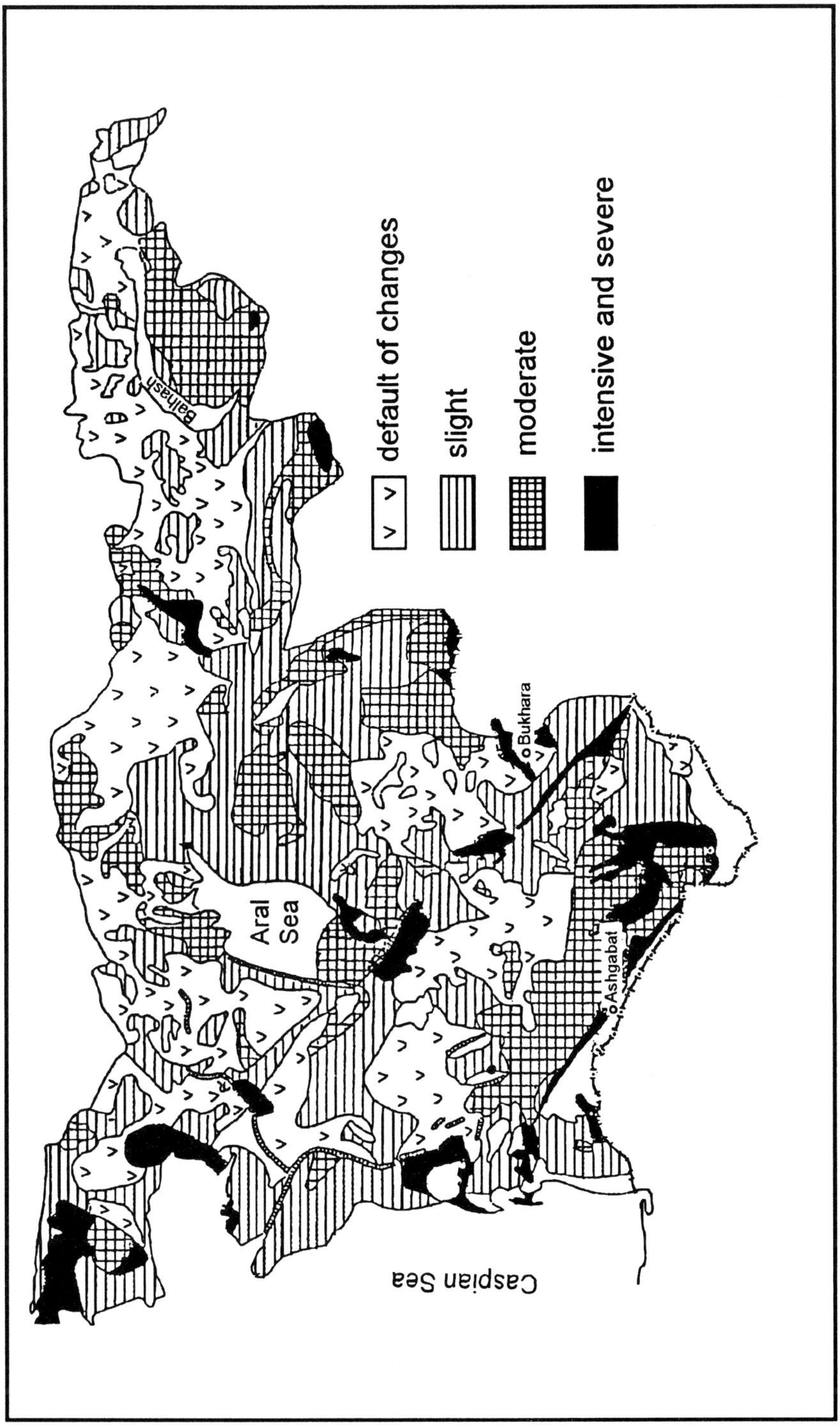

Fig. 6.3. Desertification rates of Central Asian arid zones

Table 6.4. Criteria of technological processes evaluation

Aspects	Criteria	Desertification status slight	moderate	intensive	severe	Source
PS	Reduction of veg. cover:					
	a) tree shrub plantation felling [% total area]	< 25	25-50	50-70	> 70	SP, AP
	b) reduction of turf [% of turfed area]	< 25	25-50	50-70	> 70	AP, FI
	% area of erosion, resulting from uncontrolled traffic	< 10	10-25	50-70	> 70	SP, AP
	% area covered by technological sands					
DR	Increase of areas with technological breaches (average % per year for last 5 years)	< 2	2-5	5-10	> 10	SP, AP, TM
ROD	Soil group	silt, clay, heavy loam	Loam (light and average)	supes	sand	FI, LIT
	% area of road building and other technological constructions	< 10	10-30	30-50	> 50	ST

Thus, a study of the dynamics of desertification processes and a forecast of development make up one of the main sections of regional monitoring. For these purposes it is advisable to develop such types of monitoring as paleomonitoring, retrospective monitoring, and monitoring of the present state.

The territory of Central Asia is well provided with specialized stations and stationary bases. More than 50 such stations were established where permanent observations about the dynamics of ecosystems were and are still carried out. Furthermore, about 40 nature reserves and natural preserves are in this region, enabling systematic observation of the natural environment. In this way it has become possible to set up a monitoring system to study desertification processes without requiring additional funds. But analysis of the location of this system of integrated, specialized stations and stationary bases has shown that the network of stations available at present is distributed over the territory very unevenly, and so does not reflect the current state of arid ecosystems.

In organizational terms regional monitoring of desertification on the territory of Central Asia makes it necessary to set up a subregional center (SRC) that would undertake general guidance and coordination of the work of national centers (NC), and would analyze and pool the already available and newly acquired information for the whole subregion and create a single databank.

National centers (NC) have their own base stations (BS) that carry out direct observations in nature reserves, stationary bases, permanent test grounds and key sites. Even coverage of all physiographical peculiarities of the territory and correct location of test grounds and key sites are an important organizational basis of a monitoring system.

Tab. 6.5. Criteria of soil salinisation evaluation

Aspects	Criteria	Desertification status			
		Slight	moderate	intensive	severe
PS	Salinisation rate:				
	dense [%]	0,21-0,40	0,40-0,60	0,60-1,00	> 1,00
	including Cl	0,001-0,03 (0,06)	0,030 (0,060)-0,100	0,100-0,230	> 0,230
	including Na	0,023-0,046	0,046-0,092	0,092-0,184	> 0,184
	sum of toxic salts (%)	< 0,28	0,28-0,40	0,40-0,69	> 0,69
	Distribution of salt horizon in soil profile	salts 80-100 cm	salts 30-80 cm	salts 0-30 cm	profile full of salts
	Morphological expression of salinisation, new formations	Externally salts are not expressed	Rare petty dots or stots of salts in upper dry part of profile or in loam horizon	Frequent petty dots, mould and coating in layer of 0-30 cm and their accumulation in loamy horizons	Salt mass in the shape of nests, mould, spots, dots, crystals along the profile or salty crust
	Decrease in cotton production	< 15%	15-40%	40-80%	> 80%
	Salinisated area	< 5%	5-20%	20-50%	> 50%
	Density of cotton bushes	80-50.000 per ha	50-30.000 per ha	30-10.000 per ha	< 10.000 per ha
	High of cotton at harvest period	120-80 cm	80-40 cm	40-20 cm	< 20 cm
DR	Annual increase of salinisated area increase	< 1%	2%	5%	> 5%
	Annual decrease of agricultural	< 2%	2-4%	4-8%	> 8%
	Season accumulation of salts 0-60	< 0,11-0,20% 16-30 t/ha	0,20-0,30% 30-45 t/ha	0,30-0,60% 45-90 t/ha	> 0,6% 90 t/ha
ROD	depth of ground waters	500-300 cm	300-100 cm	100-50 cm	< 50 cm
	Mineralisation of ground waters	3-6 g/l	6-10 g/l	10-30 g/l	> 30 g/l
	Drainage of aeration zones' grounds	average	slight	low	very low
	Relief of district	Plains and hilly plains	Flat plains	Hollows and channels	Hollows, channels, depressions, intercone lowering

Thus provided information should be reliable, purposeful, cover all regions and get information on all changes in geosystems. Test grounds and key sites for monitoring of desertification should be located proceeding from the optimal principles of natural zoning and zoning of a territory by the impact of human activity on the natural environment. Test grounds should be organized in the various types of physiographical regions, and key sites in main kinds of landscapes. Below a location scheme of structural units in a system of desertification monitoring in Turkmenistan is to be found (Fig. 6.4.).

The availability of vast expanses, difficult access and the severe, climatic conditions in deserts make it necessary to widely apply aerial photography in regional monitoring. Aerial monitoring of desertification processes comprises three, mutually complementing stages: space surveys, subsatellite aerial surveys and overground control. Each of these stages is required to pursue its own objectives, and should have appropriate technical, methodological and meteorological support.

6.4 Monitoring Methods

Space surveys are carried out from various technical systems installed on space carriers (photocameras, infra-red systems, side scanning system, receivers for multispectral surveys, and others). The same systems can be installed on flying vehicles (planes, helicopters).

Notwithstanding the great diversity of remote detectors, photographic materials i.e. space cameras producing photographs are more widely used for studies of desertification processes than other kinds of equipment. No other kind of space survey is so informative as space photography.

With the help of materials from space surveys it becomes possible to study, first of all, global and regional changes in the natural environment. Space photographs and images that contain generalized images of natural objects can be successfully applied in studies of quickly-moving desertification processes that develop over vast territories of deserts.

The territory of Central Asia is covered by multiple surveys from the artificial satellite "Meteor-Priroda", the orbital station "Salyut 4, 6, 7", "Mir", from spaceships "Soyuz-12" and "Soyuz-22". In recent times space information has been obtained from the first, ecological satellites "Cosmos-1906" and "Resurs-F". Images made from them are characterized by multiple repetition and high visibility, a large volume of quantitative and qualitative parameters on various objects on the Earth's surface, which is especially important in studies of dynamics of desertification processes.

Multipurpose use of information on the natural environment obtained in the course of multiple surveys from "Landsat", "SPOT" and "IRS-IA" made with the

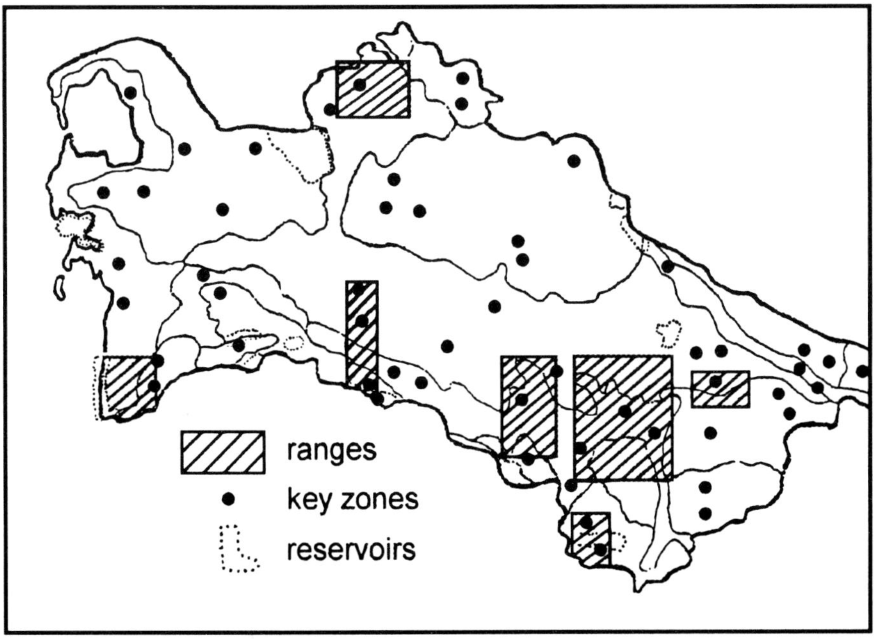

Fig. 6.4. Chart of structural units of desertification monitorig of Turkmenistan territory

help of different systems in various spectral banks has revealed their high quality and reliability. At present wide prospects are opened for obtaining and using materials from the above satellites for monitoring of desertification in Central Asian countries (Plate 6.1. and 6.2.).

Subsatellite aircraft surveys can help with understanding of many problems connected with studies of local desertification processes that are not resolved in the first stage of aerial/space monitoring.

Aerial survey is one of the best developed types of remote surveys. In spite of the fact that in recent years many more operative and not so costly surveys have appeared, aerial surveys have not become less popular. On the contrary, application of aerial surveys in combination with other kinds of surveys allow a wider and deeper study of objects on the Earth's surface.

Aerial surveys covered many key sites located in different regions of Central Asia (Plate 6.3. to 6.7.). On the basis of the use of air/space photos, various thematic maps (of desertification, vegetation, soils, pastures, dynamics of geosystems, maps of wind erosion and others) on the territory of Turkmenistan, Central Asia have been made.

Monitoring of local desertification processes is possible only if different devices are installed on one plane. In this way synchronism of all kinds of aerial surveys will be ensured, and it will automatically eliminate the influence of differences in survey conditions which is unavoidable in surveys made from different planes and at different time. This problem is solved by the development of multipurpose flying laboratories. They are created on the basis of AN-30 planes and others. Such planes can carry devices for surveys in optical, infra-red and superhigh frequency tapes, instruments for collection and interpretation of geographical information, etc.

Ground-based control is necessary to check and verify survey information obtained as a result of use of aerial/space survey materials. In some cases this stage of monitoring can resolve goals that, for some reasons, cannot be carried out in the first two stages.

Unusual, environmental conditions in Central Asia bring about certain requirements to aerial/space surveys for monitoring desertification. They depend on the nature and forms of changes made by people, poor resistance of desert geosystems to various impacts, duration of various desertification processes, low values of some desertification indices (vegetation, etc.). In this context a choice of optimal parameters of aerial surveys, in particular a spectral band, spatial resolution and an interval of repeated surveys are critical for desertification monitoring.

In aerial/space monitoring of dynamics of desertification processes the most informative zones are red and infra-red bands (600-700 and 750-850 nm). For studies of soil salinity and waterlogging most suitable are green and red spectrum bands (500-500 and 700-800 nm). The optimal, spectral bands for studies of the main indicator of desertification, desert vegetation are 400-500 nm and 600-675 nm; for studies of water erosion, the infra-red band (700-1100 nm) is used.

Taking into consideration small values of indicators of desertification processes, special attention in surveys should be drawn to indices of spatial resolution. Zoning and medium-scale mapping requires images of high resolution (30-50 m), and detailed studies of processes at least 5-10 m.

The intensive nature of desertification processes highlights the importance of the correct interval between repeated surveys. For monitoring of regions where desertification processes develop weakly, the interval of repeated aerial/space surveys is 10 to 15 years. In regions where desertification activity is moderate the surveys should be repeated every 5 or 10 years. In regions with very dynamic desertification processes, the surveys should be repeated every 3-5 years.

Account of these parameters in aerial/space surveys can enhance efficiency of desertification monitoring. In monitoring the desertification processes very important for this purpose are thematic maps made by space photographs. These maps contain diverse information on the state of components of the natural environment, including information on their degradation.

However, intensive development of desertification processes requires constant collection, analysis and storage of data necessary for map updating. An increase of the volume of such information makes it necessary to use geographical information systems (GIS) that can provide a complete, dynamic picture of natural conditions, resources, problems of nature management and conservation in a region within the confines of this territory. GIS ensures accumulation, storage and analysis of the whole mass of spatially arranged data obtained from different

Table 6.6. Periodicity of observations for desertification monitoring

Parameters	Once a month	Once a quarter	Once a year
1. Yields of pastures		+	
2. Cattle stock and load on pastures			+
3. Sand erosion			+
4. Soil salinization			+
5. Changes in land use			+
6. Water pollution	+		
7. Natural calamities	+		
8. Socio-economic indices:			
a) market prices of main products	+		
b) seasonal and annual migrations of population		+	+
c) epidemics	+		

sources. Therefore, it is necessary to establish national GIS in Central Asian countries that should be integrated into the Regional Databank of Central Asia, and later on connected to other international centers.

GIS can provide prompt information necessary for decision-makers for planning actions to combat desertification. Such GISs have already been created and operated at the Desert Institute, and we have amassed great practical experience and materials on desertification monitoring. Monitoring of desertification should include observations about one and the same parameters applying standard methods. The following minimum set of data is necessary for monitoring:

- climatic indices: albedo, precipitations, air temperature, air humidity, wind speed, and dust storms;
- description of soil and water resources: wind erosion, soil salinity, waterlogging, soil compaction, groundwaters, and surface waters;
- description of vegetation: variety composition of vegetation, biomass of tree plants, and forage resources;
- description of land use: land use system, a system of land ownership, changes in land use, a structure of sowing lands, yields of major crops, cattle stock, and loads on pastures;
- study and determination of probability of natural calamities and their prediction: mudflows, floods, severe frosts, dust droughts, etc.;
- socio-economic indices: population, dynamics of population, seasonal and annual migrations of population, mortality rates of infants, children and adults, life expectancy, epidemiological statistics, per capita income of population, income distribution, sources of income, market prices of main goods, accessible energy and prices.

As can be seen above, a system of desertification monitoring covers a wide scope of questions which can be studied applying various principles and methods, using data from various organizations and services and the establishment of new structures. Information accumulated by this time should be subject to additional interpretation, meet the interests of monitoring of desertification and nature protection. A specific feature of the monitoring system is that all its subdivisions are working under a single program.

In the monitoring of desertification, periodicity of observations is very important for operative information. To meet this requirement it is necessary to make space surveys of the whole territory once every 3 years, aerial survey of key sites every 5 years, and to prepare a review map of desertification at a scale 1 : 250,000. Maps are plotted anew every 5-10 years. Other indices of desertification require different periodicity of observations (Table 6.6).

For the operation of a system of desertification monitoring, the operative dissemination of the obtained information for decision-making and notification of the population is very important. So, it is advisable to publish a "*Bulletin of early warning*" once a quarter. It should contain data on weather, state of pastures, cattle stock, natural calamities, accidents, a death rate and birthrate of the population, etc.

6.5 Forecast

The forecast of desertification processes is an inseparable part of monitoring and requires a study of dynamics of processes in time and space affected both by natural and anthropogenic factors. For forecasts of desertification it is necessary to use and analyze materials of various scientific branches, mostly remote sensing methods that present repeated data with various time intervals. Such an approach to this problem significantly widens the volume of information and ensures completeness and reliability of desertification forecast.

The main principle of desertification forecast is a combined analysis of natural and anthropogenic factors. Methodologically it is connected with the general principles of geographical forecast. According to (Saushkin 1976) these principles can be grouped as follows:

- historical or genetic principle;
- principles of analysis of germ forms;
- principle of comparisons and analogies;
- principle of stability of historical process development;
- principle of association, i.e. forecast of a given object in interaction with others;
- principle of uncertainty or multiplicity;
- principle of incorrect forecast.

The results of historical, aerial/space monitoring of desertification comply completely with the above principles. They help understand the regularities of interactions in nature, the nature of cause-and-affect changes, determine orientation and rate of processes, and identify scales of process development.

However, it should be said that only taking into account these indices can lead to certain errors in a forecast. That is why a forecast should also take into consideration conditions of socio-economic development of the region in question, information being at the disposal of regional and local planning organizations.

Applying the above principles and materials, the Desert Institute developed forecasts of desertification processes in Turkmenistan for the next two or three decades and made a forecast map (Kharin 1983). This forecast was prepared on the basis of the planned economy of the former USSR. Later, after the transition to a market economy, this forecast needed amendment because the expected trend of desertification processes did not take place. In present-day conditions, prediction of desertification processes is complicated by some factors: a lack of stability or even crisis of some Central Asian countries and their unsteady, socio-economic development, the inadequacy of the environmental, conservation legislation, inadequate attention to ecological problems and the non-implementation of developed strategies and plans of actions to combat desertification.

Such situation leads to a still worse, general ecological situation in the region and activation of desertification processes. On the other hand, the lack of clear understanding of the mechanism of climate variations as the principal natural factor of desertification make forecasting still more complicated. In the opinion of many scientists, the climate in the next century will not be conducive to weakening of desertification processes. The awaited aridisation of the climate due to the natural course of events and anthropogenic activity (e.g. doubling of a CO_2 level) will lead to an increase of global air temperatures by 1.5-4.5°C and diminishment of annual precipitations. Thus, in the Central Asian region this will result in twice as large evaporation. All this indicates the likelihood of the acceleration of desertification processes in the future.

At present, wide-scale plans of socio-economic development of Central Asian countries are outlined, and this will undoubtedly result in a greater anthropogenic load on the natural environment. Therefore, it will be necessary to prepare and implement scientific programs: first of all for desertification monitoring in order to get new data adequately depicting the existing situation after which it should be analyzed and generalized.

References

Kharin NG (1983) Methodical principles of study mapping of desertification processes – on the sample of arid territories of Turkmenistan. Ashgabat.

Saushkin YG (1976) The history and methodology of geographical science. Publ. MGU.

7. Genetic Types of Deserts in Central Asia

A.P. Jumashov

ABSTRACT: Factors determining desert formation e.g. substrate and wind shadows are enumerated. The authors adopt an integrated approach to desert classification; the work of the Desert institute has made progress to the goal of establishing the common factors behind the genesis of deserts in which geomorphology plays a key role. The Kopetdag piedmont plain demonstrates very well the affect of sand and water in migration flow. A clear taxonomy is adopted and related to Central Asian geography: sandy, sandy-gravel and gravel, crushed stone-gypsum, loess, clay and solonchak.

For rational management of natural resources, it is necessary to define landscape-genetic types of deserts. This will ensure an all-round appraisal of natural conditions.

Geomorphological factors, in particular those that most actively influence the evolution and behavior of desert landscapes, play an important role in the shaping of various landscape-genetic types of deserts in Central Asia. They are mostly discernible in the composition of a soil profile that not only determines filtration properties, salinity, aeration, temperature variations in terrestrial air, but influences all components of a landscape.

The critical role of a substrate in the formation of microclimatic, hydrological and hydrogeological characteristics as well as soil and vegetation cover is most explicitly seen in landscape-genetic types of deserts. Consequently, the composition of deposits is one of the critical factors influencing differentiation and distribution of desert landscapes. Geomorphological conditions influence the shaping of desert landscapes also through the distribution of macrorelief forms. This refers both to eolian flows and also to water flows (surface and, to a great extent, ground). This entails, first of all, permanent displacements over many centuries involving, irrespective of an acting agent, mechanical particles and movement of salts and waters. So, in short, a macrorelief affects structures of migration flows, their direction, areas of drift, transit and accumulation of landscapes.

One very vivid example of such an influence is the landscape structure of the Kopetdag piedmont plain. Moving from mountains to a plain the following sequence of desert types is observed: loess, clay and sandy deserts. Such a sequence is a direct result of the action of a migration flow directed from mountains to a plain. A migration flow is a flow of mechanical particles, dissoluble salts and waters, both surface and ground. All these factors make their contribution to differentiation of desert types in a piedmont plain. A permanent feature of a piedmont landscape that can be seen through the whole plain shows how powerful a geomorphological factor is.

Another example of the effect of a macrorelief on the shaping of desert types is the well-known effect of a wind shadow. In such situations forms of macrorelief become obstacles in the way of a multi-century wind-sand flow. They affect the distribution of various relief types of sands featuring different fixation of substrate. If on a windward side of prevailing winds, poorly fixed, barkhan sands and averagely fixed, honeycomb-hillocky sands covered with psammophyte-*Artemisial*-shrub vegetation are developed, then on a lee side sandy plains much better fixed and, consequently, overgrown with less psammophyte vegetation will spread.

Moreover, a macrorelief together with the geological structure determine to a great extent such an important peculiarity of desert landscapes as the depth of groundwater occurrence. On the territory of lowlands, a capillary fringe often reaches the surface, or waters occur within a depth range that is accessible for plant roots.

We have adopted an integrated approach to the classification of Central Asian deserts. The account of different factors participating in desert formation makes it possible to provide an integrated appraisal of natural complexes in the arid zone, which is important for the formulation of theoretical and practical tasks.

Proceeding from a general integrated approach to the development of desert classification, first of all deserts are classified by a lithological parameter – a prevailing composition of relief-forming rocks. This is connected with the fact that different characteristics of a desert landscape often depend on this parameter.

Investigations conducted by the Desert Institute have shown that some genetic types of deserts soil, vegetation cover and other components of a landscape typical for them are characterized by specific morphology, genesis and age of a relief. This conforms to a common objective of an integrated characterization of desert landscapes. Taking into consideration specific features of general lithological-edaphic conditions in formation of Central Asian deserts, we identified in this region 6 main types of deserts (see Fig. 7.1.): sandy, sandy-gravel and gravel, crushed stone-gypsum, loess, clay and solonchak.

7.1 Sandy Deserts

Sandy deserts are a widespread type of landscape in the territory of Central Asia. They cover the Karakum, Kizylkum, Near-Aral Karakum, Muyunkum, sandy

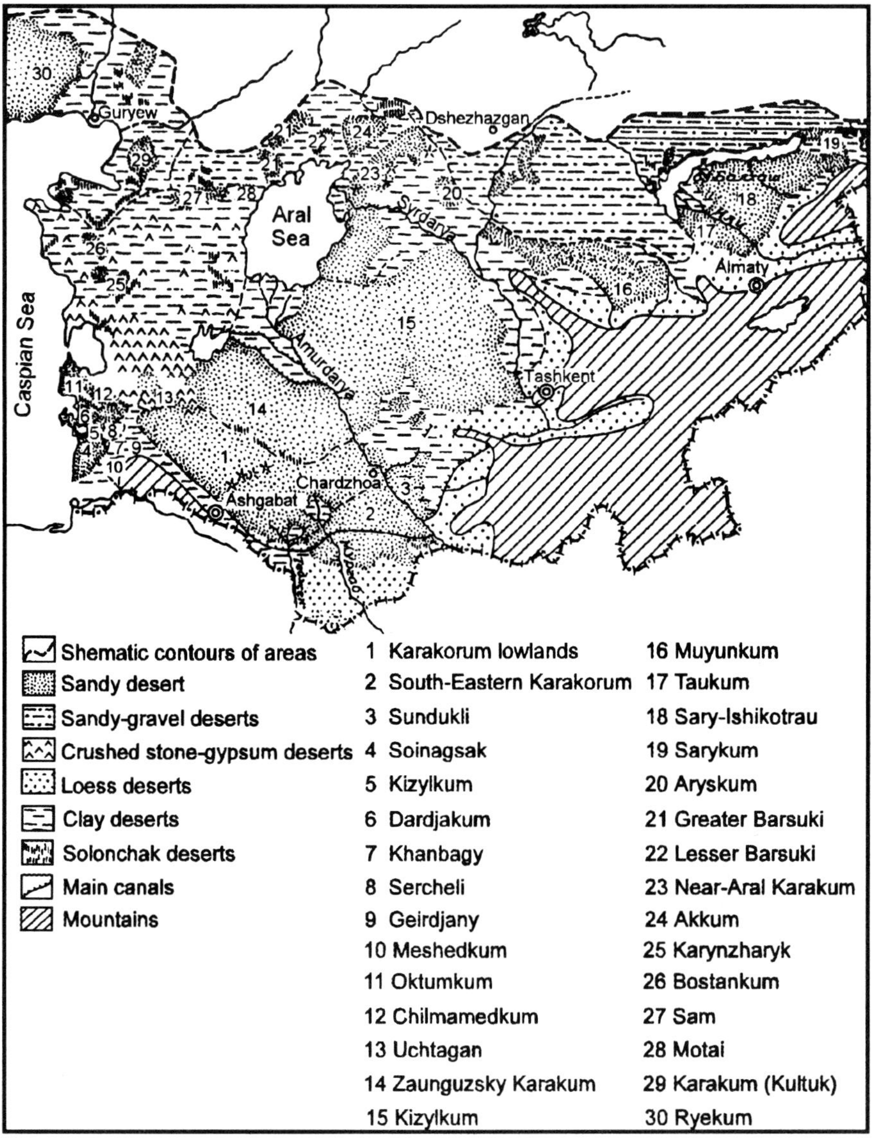

Fig. 7.1. Deserts of Central Asia

deserts in the Ferghana Depression, etc. The total area of sandy deserts is about 618,000 km² or 17% of the territory of Central Asia (Table 7.1.). These sandy deserts have a different genesis but they are developed largely over thick, loose, sandy sediments, mostly of a river and marine origin. By origin of sediments they are subdivided into the following subtypes:

- sands of ancient alluvial plains: Karakum lowlands, Chilmamedkumy, Sundukli, Kumcebshen, Kattakum, sands of the Central Ferghana, Near-Aral Karakum, Muyunkum and others.
- sands of marine lowlands: incorporate sands formed as a result of deflation of ancient and recent marine deposits. They are locally developed mostly on the eastern coast of the Caspian Sea (Western Turkmenistan).
- sands of piedmont plains: they spread mainly as separate, large, sand massifs on the piedmont plains of Kopetdag, Nuratau, Karatau and others. In their geological structure the proluvial deposits (sands, loamy sands, loams) dominate.
- sands of ancient structural plains composed of Paleogene, Neogene and Cretaceous sandstones occur mainly in the Zaunguzsky Karakum, Uchtagan, Kizylkum. Closer to the surface these plains are composed of mostly Neogene sandstones striking horizontally. More seldom outcrops of Paleogene and Upper Cretaceous sandstones and limestones are observed on large areas. Sands of this subtype are widespread.

Concerning sand mobility and thickness of a vegetation cover, the sandy deserts are subdivided into three groups:

- *Bare sands* are represented by single and group barkhans, complexes of barkhans, barkhan chains, barkhan-ridge sands. They expand over an area of more than 3 million ha, or about 7% of the area of sandy deserts (Plate 7.1.). Such sands occur largely in Western Turkmenistan, Djillikume, along the Amudarya, in the eastern part of the Aral region, in the Ferghana valley, etc. Barkhan sands also surround settlements, pits, or develop along point and linear objects. Sand formation is connected with a number of physiographical and anthropogenic factors. Affected by intensive winds, they can move as a result of which the operation of some economic units becomes impeded.
- *Semi-overgrown sands*. On the territory of Central Asia semi-overgrown sands comprise some varieties: ridge, barkhan-mound, ridge-mound, cellular, ridge-basin, and ridge-honeycomb sands. In general, they do not move. Barkhan-like mobile forms are found only on tops of ridges with a very thin, vegetation cover. Some mobile forms appear among overgrown sands as a result of an anthropogenic impact. After the end of that impact, vegetation continues growing. Barkhan sands are overgrown with *Carex physodes*, *Stipagrostis karelinii* and other pioneer grasses, then shrubs, grasses – ephemerals and still later – semishrubs. But natural, vegetative regeneration of barkhan sands is a very slow process which is connected with high mobility of sands.
- *Overgrown sands*. This group incorporates stable, often unmovable forms of a sand relief – mound, vegetated sand dunes, honeycomb and slightly rolling sands well fixed with vegetation (Plate 7.2.). After destruction of a vegetation cover, they turn within a short time period into drifting, barkhan forms.

Table 7.1. Area covered with deserts in Central Asia (Zakirov 1980)

Country	Total area	Area covered with desert		
		all types	sandy desert	%
Uzbekistan	449,000 km²	250,000 km²	107,000 km²	42,8
Turkmenistan	488,000 km²	387,000 km²	260,000 km²	67,2
Tajikistan	143,000 km²	25,000 km²	5,000 km²	20
Kyrgyzstan	198,000 km²	70,000 km²		0
Kazakstan	2,715,000 km²	747,000 km²	246,000 km²	32,9
TOTAL:	3,994,000 km²	1,479,000 km²	618,000 km²	41,8

7.2 Sandy-Gravel Deserts

This type of desert spreads mainly in the Zaunguzsky Karakum and Kizylkum. They are confined to ancient accumulative or denudation plains the surface of which was formed over sandstones, conglomerates and pebbles of the Neogene and more ancient periods.

Here one can meet small hillocks of overgrown or semi-overgrown sands. In the Zaunguzsky Karakum there are developed on meriodinally, extending ridges – kyrs, as well as in deep inter-kyr hollows of different sizes. Sometimes they can be 20-30 km long. In between large kyrs, one can see small sand relief forms (honeycomb and cone-like hollows and others). They all occur at different levels which creates an impression of a terraced relief. Their width is different varying from 100-200 m to 1-5 km. On the surface they have a sandy, often mound-shaped form with barkhans in some places.

The Zaunguzsky Karakum is composed of gray-brown or in places sandy desert soils. The vegetation cover is represented by psammophytes and halophytes, or occasionally *Holoxylon aphyllum, H.* some varieties of *Calligonum rubens, C. Setosum, Salsola arbuscula pall* and *Artemisia kemrudica, A. turanica Krasch.* Grass vegetation is poorly developed or is nonexistent at all. On sandy ridges – kyrs, for instance nearby the collective farm Damly – one can even find *Stipagrostis karelini*. In depressions there is no *Tortula desertorum*.

In the Kyzylkum the sandy-gravel deserts are formed largely from nearby residual uplands occurring in isolation or in groups. They make an area of flow formation, and removal of debris material comprising loose Cretaceous, Paleogene and Neogene deposits. These rocks form nearly horizontal layers that often outcrop to the surface or underlie others. They are often overlain by sand layers of various thickness, having yellowish-grayish coloring and an inhomogeneous composition.

Many residual uplands are broken by arch-like valleys and deep erosion gullies. Valleys are characterized by the presence of wide, flat bottoms covered by crushed stone-sandy material. Residual uplands are separated by sandy massifs. Eolian sands are sometimes underlain by a gypsum bed from 20-30 cm to 1-4 m thick. The top of the sandstones, conglomerates and pebbles has in some places sand admixtures or interbeds with significant massifs of eolian sands. Eolian forms of mound, vegetated sand dunes, honeycomb and slightly rolling sands are largely the products of weathering and deflation of the underlying Pliocene-Lower Quarternary deposits.

7.3 Gravel Deserts

Gravel deserts are met in the pediments areas and in river valleys. They make up debris cones of mountain rivers. In the upper reaches of rivers, they are represented by large boulders and pebble of different sizes, in the lower reaches by gravel and sands. Thickness of interbeds is not the same. In some places the thickness of pebble deposits reaches 10 m. A surface of piedmont plains is usually slightly broken; in some places the traces of intermittent waterstreams are visible; gullies are filled with loose, mostly loamy sand and clay material.

Groundwaters are low saline and fresh, they occur at a depth of 4-8 m. Soils are poorly developed, loamy sandy, often overlain by pebble, used as pastures and partially for farming. Density of the natural vegetation cover of sandy-pebble Kyzylkum deserts increases with the growing thickness of sands. It is represented by *Haloxylon persicum*, *Caliginous rubbers*, *Tamari androssowii*, *T. raamossima*, *Artemisia kemrudica*, *Astragalus paucijugus*, *Salsola arbuscula*, *Ferula litvinoviana*, *Salsola orientalis*, *Astragalus longitiolatus* and others.

7.4 Crushed Stone-Gypsum Deserts

These are developed in the north-western part of the Krasnovodsk plateau, on the Tuarkyr, Kanlankyr, Ustyurt, Mangyshlak, in the Kyzylkum on sheet Paleogene and Neogene plateaus, in Betpakdel and other regions. Spreading over a vast territory they cover different relief forms – structurally denuded lowlands; denudation plateaus; alluvial-proluvial, upland, piedmont plains, the surfaces of which are complicated to different extents by recent, relief-forming processes.

For instance, on the plains of Mangyshlak and Ustyurt there are many closed depressions, mostly of karst origin (wide, valley-like depressions devoid of any vegetation). Karst cones occur mostly nearby escarpments of depressions (Borovskyi 1974). A surface is composed largely of limestones, sandstones, crushed stone, conglomerates, clays over which gray-brown gypsum soils are formed.

Gypsum occurs at a depth of 5-30 cm, and has a brown, gray, reddish and other coloring. Its content in soil reaches 50-80%. In some areas of Ustyurt apart from crushed, stone-gypsiferous genetic types clay and at times stony types are present. A surface of crushed, stone-gypsum deserts is practically devoid of eolian forms of relief, and only in some places, under shrubs very small fixed sand dunes up to 0.3-0.5 m high are met which do not form anywhere their own continuous massif.

A river network within the crushed stone-gypsum desert is developed poorly, unevenly. It is represented by small rivers originating in springs in a piedmont zone. Riverbeds of temporary, rainstorm streams are widely developed. The waters of the rivers Sarysu, Mointi, Tentek and others is used for irrigation and water supply. Groundwaters are mostly saline and occur at a depth of 10-60 m.

Natural vegetation of deserts is represented by communities of low semishrubs. Depending on soil texture and thickness of a crushed, stone-gypsum series, composition of the vegetation cover is changing, too. On loamy and loamy sands the most often met are *Nabasis brachiate*, *Eremopyrum bonaepartis*, *Convolvuens divaricatus*, *Istpatranscoucosisa grosth*, *Poa bulbosa*, lichen and ephemerals. On clay and loamy soils *Salsola gemmascents*, together with *Anaphyton erinaceum* and ephemerals grow. On crushed stone soils, communities of *Anabasis brachiata* with some varieties of Artemisial may be noted.

In crushed, stone-gypsum deserts, gray-brown and brown soils are widespread, as well as complexes of brown, solonetzic soils with solonetz. Takyr and takyr-like soils are met only in valley-like depressions, have a deluvial origin and are distinguished by weak, gypsum accumulation and relatively low salinity (Lobova and Shuvalov 1980).

7.5 Loess Deserts

Loess deserts occupy the central and southern parts of Central Asia, and largely cover a transitional zone between the mountain systems of Tien-Shan, Pamir-Altai and the enormous, sand expanses of the Karakum and Kyzylkum (Plate 7.3.). Vast loess areas are confined to the Pritashkensky region, the Golodnaya steppe, the Ferghana depression, the Samarkand lowland, Kitabo-Shakhrisyabzsky inter-mountain depressions, the Kopetdag piedmont plain, the Southern-Tajik depression and other regions.

The geomorphologic range of loess development is considerable. Loess soils occur in river terraces, piedmont plains, uplands and plateaus.

Loess soils in Central Asia have a different genesis and age. According to G.A. Mavlyanov's classification (Mavlyanov 1958), taking into account variations of local conditions of loess distribution, in its composition the following genetic types can be singled out: eolian, proluvial, deluvial and alluvial genesis.

Their age varies from the Lower Quarternary to the modern. Loesses and loess-like formation of the Middle and Upper Quarternary age are widely developed. Thickness of deposits ranges from 2-3 m to 100 m (Mavlyanov *et. al.*). Loesses are most thick in piedmont plains and uplands.

Eolian loess is developed largely on piedmont plains of the Aktau mountains, the Charkalsky range, Kopetdag and others. It occurs on the Neogene, Paleogene and more ancient deposits often covering them with a continuous layer up to 2-8 m thick. The deposits are characterized by a yellowish-gray coloring, good sorting of the material, and lack of bedding.

Deluvial loesses and loess-like deposits are widespread in mountains and piedmonts of the Gissar, Zarafshan and Turkestan ranges. They are largely made up of products of weathering of parent rocks and seldom of re-deposited, alluvial psephitic and small-grained, loose deposits. Deluvial loesses and loess-like deposits are not bedded, compact. Their thickness ranges from 1-2 m to 10-15 m.

Alluvial loess deposits are widely developed in inter-cone depressions, in deltas and valleys of mountain and other rivers (Chacha, Kashan, Sokh, Chirchik, Zarafshan, Kashkadarya, Surkhandarya, Syrdarya and Amudarya). They are formed by weathering and washing of loess, loess-like deposits and other mountain rocks. They are of a yellowish-gray coloring, bedded, with lenses of sand and gravel, often underlain by loose soils and pebbles of a river origin. Debris cones of the Kopetdag and Nurat ranges are composed largely of coarse textured series, while inter-cone depressions and marginal sites have medium- and fine-grained sediments. Inter-cone depressions and peripheral parts of debris cones feature a smooth relief.

Proluvial loesses and loess-like deposits are made up of products of weathering of parent rocks, redeposited by waterstreams and eolian processes. They have mostly a grayish-brown coloring, sometimes they have different tints due to admixtures of products of weathering of mountain rocks in a water intake basin. Their thickness in the Pritashkent region is 40 m, the Samarkand depression 35 m, the Karshi steppe and Kopetdag piedmont plain 30 m. The greatest width of a zone of these deposits occurrence in the Golodnaya steppe reaches 150 km (Mavlyanov *et. al.*).

Proluvial loess deposits are mostly layered. Interbeds of clays, sands, gravel and crushed stone are observed in them. Due to a loose structure and considerable thickness, they feature high porosity. Moving downwards the size of particles increases because in the upper part of alluvial loess deposits, the eolian sediments participate in the complex. Their thickness is not more than 6-10 m. Their thickness and composition vary with an increase of remoteness from the mountains. Increases of thickness of loess deposits is connected with the age of terraces i.e. the younger a terrace, the less is the thickness of a loess cover. In general, these deposits are non-subsiding.

Groundwaters in loess deserts occur at a great depth (20-100 m and more). They are mostly fresh or low saline. Vegetation cover is monotonous here which is connected with soil properties, hydrological conditions, relief and climate on the territory. In wet years ephemerals, large cereals and other plans grow in great quantities on light serozems. On slightly broken piedmonts are Artemisial, sedge-meadow grass associations with *Salsola dendroides*, *S. orientalis*, and *S. gemmascens* grow widely. On flat plains composed of highly saline, heavy loams and clays, annual thistle and halophytes dominate.

In loess deserts, typical and light serozems makes the best part of the land stock in Central Asia. A considerable area of loess deserts is developed for irrigated and rainfed farming. Steep slops and lands unsuitable for agricultural crop cultivation are used for grazing. The forage capacity of loess deserts depends on precipitations. In a favorable year the natural productivity of pastures in loess deserts increases considerably.

7.6 Clay Deserts

Clay deserts are to be found in South-Western Turkmenistan, on the Kopetdag piedmont plain, in ancient deltas of the rivers Amudarya, Syrdarya, Murghab, Tedjen, Kashkadarya, Zarafshan and others. They extend over areas with a flat relief, formed on clay and loam, proluvial and alluvial sediments of the Neogene and Quarternary age. Soils are represented by takyrs and takyr-like types (Plate 7.4.). Their surface is compact, polygonal-fissured, the size of polygons vary from 3-4 cm to 10-12 cm in a cross-section. They are underlain by parent rocks. Their thickness is from 0.5-10 m and more.

7.7 Takyrs

Takyrs and takyr-like plains in the Lower Karakum and Kyzylkum have a thin, water-permeable crust, while in other regions takyrs are composed of a clay layer several meters thick. Their texture is inhomogeneous. Sites of clay deserts alternate with sands or are partially overlain by eolian sands. An area of pure takyrs is not large. In clay deserts dry beds are met quite often. When precipitations fall, they turn into temporary waterstreams. Groundwaters occur at a depth of 10-20 m and deeper.

Vast areas of clay deserts are, in fact, devoid of higher vegetation. In some places annual thistle, water grass, lichen, as well as *Artemisia spec.*, *Salsola orientalis*, *S. arbuscula pall*, *S. dendroides*, *Atriplex cana* and others grow. Clay deserts are natural watershed areas. Unfortunately, their surface is broken by heavy transport vehicles, except for small spots met in some nature reserves.

In Central Asia solonchak deserts have developed nearly everywhere and are met in depressions where intensive salt accumulation is taking place in conditions of high occurrence of saline groundwaters and an arid climate. The drainless nature and poor drainability of the territory greatly affect the formation of solonchak deserts.

In a desert zone and also in oases solonchaks take up a considerable area. Large areas of solonchaks are found in deltas of the Amudarya, Syrdarya, Murghab, Tedjen, Atrek, Kashkadarya and Zarafshan rivers, in the eastern part of the Kopetdag piedmont plain, as well as in South-Western Turkmenistan. They occur widely on the coast of the Caspian and Aral seas, in the Akchakaya, Mullaly, and Aidar depressions.

7.8 Solonchak Deserts

Solonchak deserts are lifeless territories without any prospects. Solonchaks contain a great quantity of water soluble salts, and as a result there are very few, nutritive elements in them. Depending on the nature of salinization and the groundwater salinity, they are subdivided into crust, puffed, crust-puffed and wet. By origin and suitability for irrigated farming solonchaks are subdivided into primary (sors) and secondary.

Primary solonchaks cover lowered areas and are composed largely of heavy deposits with a high content of water soluble salts.

Secondary solonchaks are confined to regions of irrigated farming and appear as a result of an intensive rise of groundwaters. In oases of Central Asia meadow, solonchaks occur like islands in low elements of a relief and develop over alluvial deposits of various textures. The morphologically specific feature of meadow solonchaks is the presence of a sizeable, salt layer 3-7 m thick, having a dirty-whitish coloring. This fragile, salt crust is composed nearly completely of gypsum, carbonates and other water soluble salts.

Groundwaters in such solonchaks have a mixed regime. They are combined with seepage, drainage waters from irrigation canals and fields.

Areas of meadow solonchaks in irrigated regions grow from year to year. When in a natural state, solonchaks are covered on the surface with a loose, salt crust. That is why their development for farming is practically impossible without reclamation improvements. For instance, after landing on cultural plants during their flowering, salts blown away from solonchaks, kill pollens and thereby decrease yields. Processes of soil salinization are caused by natural and anthropogenic factors. Depending on the chemical composition, the soils can be sodic, chloride, sulfate and nitrate.

Large areas are covered by solonchaks along the coast of the Caspian Sea in Cheleken, Dardja, and around the Karabogazgol Bay. They are mostly wet with a puffy crust. Their surface is usually flat, often small and hillocky, cut by minor tributaries. They gradually merge with the surface of coastal shallows of the Caspian. In the western part of the coastal territory the inter-ridge hollows are often filled with sandy solonchaks - the former seabed. These solonchaks are covered by gypsum sands, loamy sands and loams.

The lowering of the Aral Sea level is also conducive to the active formation of solonchak deserts. This led to the drying out of more than 2,100 km^2 of the former seabed. By 1990 water in the eastern coast of the Aral Sea had receded by 100 km, and this process has continued intensively. Major changes occurred in the natural environment of the Amudarya and Syrdarya deltas: enormous inundations stopped, many shallow lakes and marshes dried out, a general drying out of soils and their surface salinisation have been observed.

Solonchaks in the dried part of the Aral Sea are largely wet. The zone of aeration contains considerable reserves of salts, thus, only halophytes (annual thistle, karabarak and others) can vegetate here, but they have no practical value for grazing of cattle.

References

Borovski VM et. al. (1974) Soils of Mangyshlak Peninsula. Alma Ata. Nauka.

Lobova EV, Shuvalov SA (1980) Survey of soil investigations on Ustyurt. Nature Soils and Problems of Ustyurt desert development. Pushchino: p. 3-8.

Mavlyanov GA (1958) Genetic types of loesses and loessics species of central and southern Central Asia. Tashkent.

Mavlyanov GA et. al. (1966) Loess species of Central Asia. Loess species of USSR. Nauka: p. 214-221.

Zakirov VM (1980) Railways in sandy deserts. M. Transport: p. 221.

8. The Mapping of Desertification Processes

T.A. Babaeva

ABSTRACT: A new cartographic method developed at the Desert Institute is described and elucidated. A new set of quantitative criteria have been worked out to assess the intensity of erosion processes. The maps contain much useful information on desertification especially the risk factors. Contrary to the usual assumptions, waterlogging was demonstrated not to be a causative factor. General mapping problems are explained. The usual maps are too local, too small scale to identify general desertification risk signals. The methods of surveying necessary for cartographic purposes will include remote survey and specialized reports. International projects (e.g. Mali and Mongolia) demonstrated the usefulness of these methods. Of especial interest is that for the Aral sea basin.

Desertification control requires targeted action especially in the countries where these processes have become catastrophic. In this connection an international project is being prepared on the assessment and mapping of desertification (FAO 1981).

In 1985 the Desert Institute perfected an estimation method and a cartography of desertification processes as a model for the territories of Central Asia, and the Institute has tested it in different natural conditions.

8.1 Map: World Deserts and Desertification

Desertification maps are considered as being complex, nature conservation maps. At present there are several types of such maps in existence, and they apply different methodological approaches. One of these maps is "*World Desert and Desertification*" with a scale of 1:25,000,000 (Centre National de la Recherche Scientifique 1976) which covers the entire territory of arid regions in the world. In spite of its schematic nature and small scale, it contains in a generalized form information on desertification processes. It depicts the following groups of factors that contribute to desertification:

- natural risk of desertification (sand massifs and drifting sands, accumulation of stones and rock debris, areas devoid of vegetation, and water erosion);
- the impact of people (territories with a high population density and places of intensive cattle grazing);
- combating desertification;
- the degree of desertification (high, medium, and low).

A risk of desertification is marked on the map by the intensity of respective twin signs. Desertification processes are characterized as follows:

- sand dunes, moving sands, sand drifts with a risk of desertification from high to medium;
- enormous accumulation of stones or a desert shield formed as a result of wind erosion with a risk of desertification from medium to low;
- surface compacting with the increase of surface runoff, decrease of accessible moisture, and poor germination of seeds.

8.2 Overview Map of Desertification in Turkmenistan

This map was used in the Desert Institute to make an overview map of desertification in Turkmenistan with the scale 1:2,500,000 (Babaev *et. al.* 1978). A risk of desertification is shown as in the above map by twin signs.

- Signs 1 and 2 in the legend denote ridge, ridge-mound and mound sands, varyingly overgrown by vegetation, with a risk of desertification from high to low.
- Sign 3 denotes accumulations of stones and a desert shield formed as a result of wind erosion: risk from medium to low.
- Signs 6 and 7 denote compacted clay surfaces, takyrs devoid of any vegetation: risks from high to medium, and from medium to low.
- Signs 8 and 9 denote saline gray-brown soils, sometimes solonchaks, devoid of vegetation: from high to medium and from medium to low.
- Signs 10 and 11 denote territories affected by water erosion.
- Sign 12 denotes territories with a high population density.
- Sign 13 denotes territories with a high concentration of cattle on pastures. Signs 12 and 13 denote territories affected by human activities. The rest represents natural landscapes.

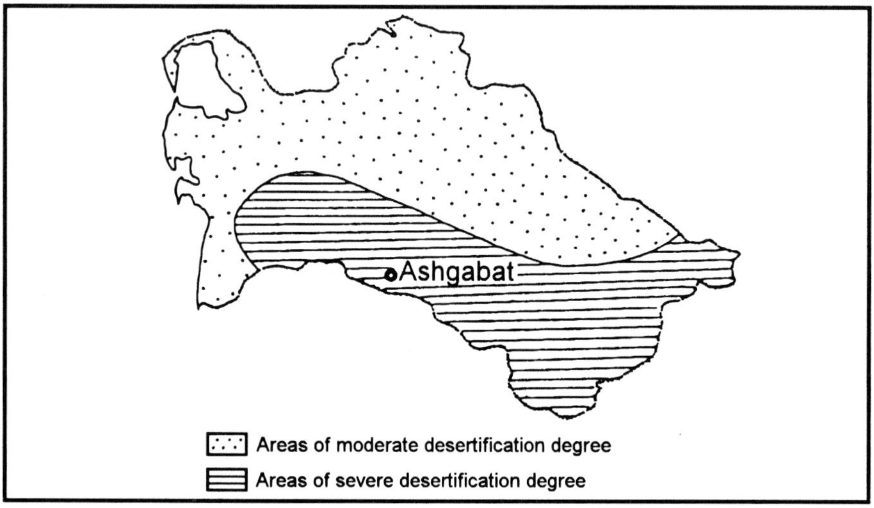

Fig. 8.1. Fragment of the „Desertification Map of the World".

8.3 Desertification Map of the World

The "*Desertification Map of the World*" (UN 1977) was prepared by the Desert Institute for the UN Conference on Desertification (UNCOD). In this map on the territory of Turkmenistan, the following bioclimatic zones are delineated: arid and semi-arid distinguished by the aridity index. Non-scale signs denote territories prone to desertification, and the latter can be divided into the following categories:

- regions where sand movement is possible. This category includes sandy deserts where a risk of wind erosion is very high;
- stone and rock surfaces where a risk of water erosion is high. A risk of desertification is shown on the map by background coloring.

The map was prepared using generalized information on the state of desertification in arid zones of the world (Dregne 1977). On the territory of Turkmenistan the following classes of desertification are shown:

- moderate desertification: territories with accelerated wind and water erosion, and places where soil salinity causes 10-50% diminishment of yields;
- severe desertification: territories with well-developed sheet erosion; territories where soil salinity causes more than 50% diminishment of yields (Fig. 8.1.).

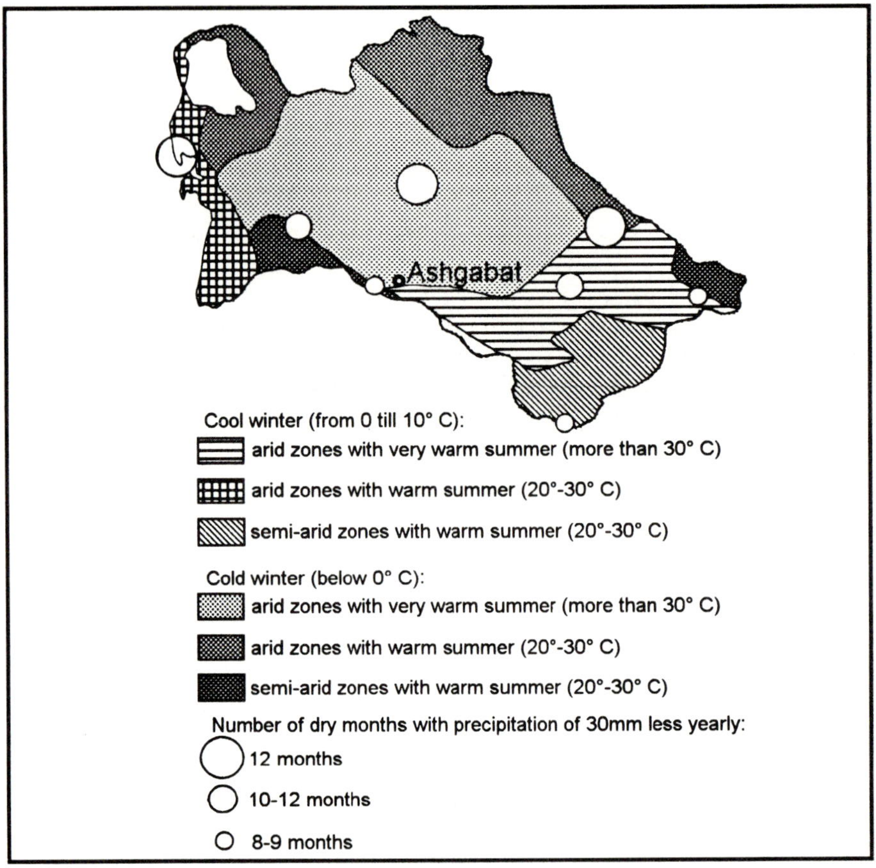

Fig. 8.2. Fragment of the „Map of World Distribution of Arid Regions".

8.4
Map of World Distribution of Arid Regions

The *"Map of World Distribution of Arid Regions"* (1977) depicts climatic conditions in the world's arid regions (Centre National de la Recherche Scientifique 1977). It shows the degree of aridity (extra-arid, arid, semi-arid and subhumid) determined by a ratio of average, annual precipitations to potential evaporation. Durations of dry periods and the accompanying precipitation are shown by background coloring. Distribution of droughts by seasons is shown by circles. A region's aridity is determined by a ratio of precipitations to evaporation (Fig. 8.2.).

8.5 Map of the Present State of Desertification in Turkmenistan

The *"Map of the Present State of Desertification in Turkmenistan"* with the scale 1:1,500,000 provides information on development of some types of desertification, the intensity of desertification processes and their causes (Desert Institute 1985).

On the map the desertification classes are shown by background coloring, while on the map in 8.1 by different hatching. Types of desertification and improvement of the natural, desert environment are denoted by non-scale signs. And last, the causes of desertification are denoted on the map by letters. The map also shows territories where irrigated farming is practiced, and where considerable changes in the natural environment occur.

8.6 Map of Desertification Risk

The *"Map of Desertification Risk"* was prepared by the Desert Institute of Turkmenistan on the same scale as the map of the present state (see 8.5). In comparison to the map of the present state, a criterion of a desertification class is replaced by a criterion of desertification risk, and one more criterion is added: the direction of desertification processes in comparison to the present state. Of all causes of desertification they include only natural and anthropogenic. The map was prepared by the results of analysis of the map of the desertification present state, and studies of space photographs and specialized literature (Plate 8.1.).

On the map the desertification risk is shown by background coloring. Non-scale signs denote types of desertification and improvement of the desert natural environment. Causes of changes and direction of the desertification process are denoted by letters.

8.7 Anthropogenic Desertification of the USSR Arid Territories

The "*Map of Anthropogenic Desertification of the USSR Arid Territories*" with a scale 1:2,500,000 was worked out by the method developed by the Desert Institute with regard to FAO and UNEP elaborations. It contains integrated information on the total risk of anthropogenic desertification (Desert Institute 1985). This index is a predictable characteristic showing tendencies in development of the desertifica-

Table 8.1. Risk of desertification in Central Asian arid lands (Babaeva 1983)

Types of desertification	Kazakstan*	Uzbekistan	Turkmenistan	TOTAL
Degradation of vegetative-cover due to overgrazing	481,513 km² 44.2%	47,127 km² 17.1%	125,547 km² 30.3%	654,187 km² 36.4%
Degradation of vegetative-cover due to undergrazing			3,758 km² 0.9%	3,758 km² 0.9%
Territories with partial desertification round wells	8,714 km² 0.8%	51,256 km² 18.6%	42,168 km² 10.1%	102,138 km² 5.6%
Wind erosion		11,024 km² 4%	48,012 km² 11.5%	59,036 km² 3.2%
Salinization of arable lands	6,536 km² 0.6%	3,993 km² 1.4%	20,875 km² 5%	314,004 km² 1.6%
Technological desertification	61,006 km² 5.6%	41,884 km² 15.2%	12,942 km² 3.1%	115,832 km² 6.5%
Salinization caused by a sea level lowering	75,162 km² 6.9%	7,993 km² 2.9%	11,690 km² 2.8%	94,845 km² 5.1%
Water erosion			12,228 km² 2.8%	12,228 km² 0.7%
Desertificated areas in total	632,931 km² 59.9%	163,277 km² 59.2%	277,220 km² 66.5%	1,073,428 km² 61.9%
Unapplied cultivation	19,609 km² 1.8%	13,507 km² 4.8%	25,468 km² 6.1%	58,584 km² 3.2%
Areas unsubjectable to desertification	436,835 km² 38.3%	98,841 km² 36%	114,812 km² 27.4%	709,072 km² 39.8%
TOTAL	1,089,375 km² 100%	275,625 km² 100 %	417,500 km² 100%	1,782,500 km² 100%

* only arid lands

tion process in the most recent period. On the map this index is denoted by coloring distinguishing very severe, severe, moderate and low desertification categories (Plate 8.2.)

The legend of the map also includes marks showing the effect of population density and cattle loads on pastures. Letters on the map denote the following aspects of desertification: present status (PS), desertification rates (DR), internal risk of desertification (ROD), impact of animals (domestic) on the natural environment (IA), anthropogenic impact (AI) determined in relation to population density. As natural prerequisites of desertification processes the map uses climatic indices (an aridity type). It follows from the map that the percentage of areas threatened by desertification is practically the same in Kazakstan and Uzbekistan (Table 8.1.). In Turkmenistan this index is somewhat higher. However, the degree of desertification risk in some republics of Central Asia varies as it is related to natural conditions, specific features of farming and the direct impact of people on the natural environment. Desertification maps of Mongolia and Mali were prepared applying the same methodology (Kharin *et. al.* 1992, Kharin *et. al.* 1993)).

In the framework of bilateral scientific cooperation between the Academy of Sciences of the USSR and the Mongolian People's Republic, the Desert Institute studied the desertification processes and worked out an action plan for the protection of nature in the arid ecological system of Mongolia. The Desert Institute took part in the USSR International project (UNEP) for Mali "Support to West Africa in the struggle against the desertification by strengthening national activity through researches and training of specialists". The aim of this project was the testing of the elaborated method of research and cartography into the desert processes in Mali.

8.8 Map of Anthropogenic Land Degradation in the Aral Sea Basin

One of the most recent desertification maps is the "*Map of Anthropogenic Land Degradation in the Aral Sea Basin*," with a scale 1:2,500,000 (Desert Institute 1993). Using a system of symbols, the map characterizes the following processes: degradation of vegetation cover, deflation, water erosion, salinization of irrigated lands, soil salinization caused by lowering of the Aral Sea level, technogenic desertification, the waterlogging of pastures in the command zone of irrigated canals, pollution of air in cities, as well as degrees and classes of degradation of geosystems (Plate 8.3.).

As there are no data on a "background level" of geosystems the first group includes undisturbed and slightly disturbed geosystems. Some categories of lands that are not used in the national economy are not included in the assessments. They are: mobile sands of natural origin, solonchaks, outcrops of parent rocks, and territories (a belt along the Turkmenistan frontier) which resemble nature reserves in their use.

8.9 Outlook

Analysis of the maps has shown that they describe desertification processes worldwide taking different methodological approaches. A common drawback of these maps is too small a scale. It follows that the next goal of desertification mapping is the preparation of maps of larger scales.

A degree of anthropogenic desertification is assessed by specially developed criteria. First of all, they identify types of desertification processes such as the degradation of a vegetation cover, wind erosion, water erosion, salinization of irrigated lands, technogenic desertification, and soil salinization.

Table 8.2. Assessment criteria of a vegetation cover degradation

Criteria	Desertification classes		
	low	moderate	high
1. Diminishment of productivity (Yield) [%]			
Sandy desert	< 15	15-35	< 35
Gypsum desert	< 20	20-40	< 40
Clay desert	< 35	35-70	< 70
Piedmont loess desert	< 20	20-40	< 40
Mountain vegetation, lower belt	< 20	20-40	< 40
Middle and upper belts	< 15	15-30	< 30
2. Diminishment of projective coverage of native vegetation [%]			
Sandy desert	< 10	10-35	< 35
Gypsum desert	< 10	10-50	< 50
Clay desert	< 20	20-45	< 45
Piedmont loess desert	< 20	20-40	< 40
Mountain vegetation, lower belt	< 10	10-50	< 50
middle and upper belts	< 5	5-40	< 40
3. State of Vegetation cover	climax or slightly changed communities	long-derivative communities	short-derivative communities

The criteria of desertification may be local, regional or zonal. Local desertification criteria describe specific features of desertification processes within separate landscapes. For instance, they can include data on a variety composition and projective coverage of particular vegetation communities. These criteria were used in preparation of large-scale desertification maps.

Regional criteria are developed for geographical regions. They contain generalized data on anthropogenic disturbances of the natural environment. For instance, while preparing desertification maps for Turkmenistan they applied criteria describing development of desertification processes in the Karakum desert.

Zonal criteria are more general. They characterize the development of desertification in the arid zone of CIS in general. In these cases diminishment of productivity in these criteria was expressed not in absolute figures, but in percent. This enabled comparison of desertification processes in different types of deserts in which vegetation productivity is expressed in absolute figures. For other arid regions, criteria can be expressed otherwise than those developed for CIS countries. For instance, for savanna in the western regions of Mali a degree of vegetation damage by fire is of great importance.

To assess the state of territories in terms of desertification one can use particular objective indices describing a state of a vegetation cover. Most typical quantitative indices are taken for different types of deserts, for instance, coverage by native vegetation (Table 8.2.).

Enhancement of desertification is attended by diminishment of the nutritive capacity of pastures, which can be attributed to a replacement of well-devoured plants with lower quality plants and an accompanying decrease of pasture yields. In fact, severely desertified territories are unsuitable for cattle or sheep grazing, and should be improved by introducing valuable, forage plants.

It is not correct to relate deflation processes only to natural causes because no tectonic dislocations are observed as a result of sharp changes of natural conditions due to climate variations. Anthropogenic processes connected with river flow regulation, and putting into use the earlier unused, vast territories remain the critical factors.

The following are taken as deflation criteria : an area of drifting sands, a non-projective layer of grass, and a ratio of projective coverage by the shrub vegetation to an area under thicker grass vegetation (Kharin *et. al.* 1993).

If all positive forms become mobile, then vegetation cannot root in them and deflation becomes very severe here. Plant turf creates a protection layer on the surface and can be taken as a criterion.

Shrub vegetation creates turbulent air flow, while grass vegetation weakens the latter near the Earth's surface. A combination of these types of vegetation make it possible to use them as a criterion.

In spite of low precipitation, other factors such as dry and hot climate, scantiness of a vegetation cover and spreading of easily scoured sedimentary deposits contribute to development of water erosion in some regions of Central Asia. Among the factors producing water erosion and its development are the regime and quantity of precipitation, relief, the watershed area, soils and vegetation.

Overgrazing causes compacting of soil by cattle, and degradation of pasture vegetation. This reduces infiltration of precipitation into the ground, increases the surface runoff, and diminishes the moisture content of the soil. As a result, water erosion becomes more intensive, while pasture vegetation suffers from inadequate moisture supply. All this leads to the dominance of xerophytic vegetation and the development of soil erosion.

The following quantitative criteria are used to assess the intensity of erosion processes: the type of water erosion, washing off of a soil top horizon, and diminished projective coverage by vegetation. Quantitative indices are set for each criterion (Kharin *et. al.* 1993). For the determination of a class of water erosion it is necessary to take into consideration two main indices: a projective coverage by vegetation and steepness of slopes. For the middle and upper belts of mountains it is necessary to take into consideration a projective coverage by tree-shrub vegetation; for the low belt, the same index, but for semi-shrub vegetation.

Irrational and unwise use of water and land resources affects natural factors causing alteration and later lowering of the biological productivity of irrigated lands. Salinization is connected with high air temperatures and intensive evaporation from soil. Groundwaters are the main source of soil salinization.

In determination of degradation intensity in an irrigated zone, the following criteria can be assumed: a degree of salinity by a solid residue, groundwater salinity, salinity of irrigation water, yields of major crops (cotton), seasonal build-up of salts and pollution of irrigation water (Table 8.3.).

Table 8.3. Evaluation criteria for arable lands salinization

Criteria	Desertification status		
	slight	moderate	intensive
1. Rate of salinization [%]:			
on dense remainder	0.21-0.4	0.4-0.60	> 0.60
including: chlorine-ion	0.001-0.03	0.03-0.10	> 0.10
including: sodium-ion	0.023-0.046	0.046-0.092	> 0.092
2. Mineralisation of sub-soil waters [g/l]	3-6	6-10	>10
3. Mineralisation of irrigating waters [g/l]	0.5-1	1-1.5	>1.5
4. Reduce of raw cotton crops [% from background level]	15	15-40	40-80
5. Season salt accumulation			
in %	0.11-0.20	0.21-0.30	0.32-0.60
in t/ha	16-30	30-45	45-90
6. Degree of irrigating water pollution	Slightly polluted	polluted	severely polluted

A correlation between desertification classes and a degree of salinity is as follows: low risk class; low salinity and middle risk class; medium salinity and high risk class; severe salinity. The latter includes solonchaks.

Salinization of a subsurface layer prevents development of the root system of cotton. Yields of raw cotton connected with soil salinity can be one more criterion of desertification.

Desertification can move forward not only in a continuous front over a vast territory, as was the case in frontier regions of the Sahara desert, but can also cover small territories and propagate locally. An apposite example here is desertification caused by technical facilities.

A technologically caused (technogenic) desertification includes that caused by technical facilities in the course of construction, survey and other works performed in a desert. Among likely criteria of this type of desertification we can name destruction of vegetation cover, erosion development, territories covered by technogenic sands, and length of roads (Kharin *et. al.* 1993).

According to our investigations, land waterlogging was not related to desertification, because during the course of waterlogging, productivity of a vegetation cover becomes higher, and that is not characteristic of desertification. However, in this case we relate this process instead to degradation of geosystems, because their stability is violated.

The given criteria of desertification assessment are not final and generally recognized in the world. They were developed mostly for arid territories of Central Asia that suffer most from degradation of a vegetation cover, deflation, water erosion and soil salinity. Degradation is observed on more than half of the territory of the region in question, while heavy degradation affects about 7% of the territory. After improvement of these criteria, they could be used as universal criteria.

For preparation of desertification maps it is necessary to have operative information that can be provided by remote surveys. Maps were prepared using different materials. First of all, these were space photographs at various scales (1:200,000-1:1,000,000) and types (panchromatic, color, synthesized, and spectrozonal).

The second important source of materials for mapping is thematic maps. Their combined analysis, and the interpretation of space photographs and specialized sources will help attain the set goal.

In the first stage of map preparation, the cartographers study available specialized sources on the state of desertification, develop a legend for a map, and quantitative criteria to assess desertification processes. In the second stage, space photographs are interpreted in laboratory conditions. Here they use as a reference, aerial photographs of the regions where surveys were carried out earlier. At this stage they outline field sites that are to be visited to check the results of laboratory interpretations. During the third step, results of laboratory interpretation of aerial-space photographs are checked under field conditions, and the earlier suggested criteria of desertification are verified. And finally, the fourth stage is reserved for checking the desertification map, making additions and alterations that were identified in the course of field work. The legend is finalized and the map is drawn.

Space photographs can help investigate local processes (e.g. spots around pits) and the dynamics of desertification processes e.g. technogenic desertification (Babaeva 1982, Babaeva 1983).

In view of the deterioration of the global ecological situation, more and more attention is being drawn to environmental protection mapping. Materials available from the Desert Institute of Turkmenistan include an item on the preparation, publication and dissemination of the Desertification Map of the World at a scale 1:5,000,000. These maps can be used in planning nature conservation actions.

References

Babaev AG *et. al.* (1978) Integrated desert development and desertification control in the Turkmenistan SSR. Environmental Sciences and Applications, Vol. 12. Pergamon Press. London.

Babaeva TA (1983) Study of desertification processes dynamics in space photographs. Probliemy osvoieniia pustyn, 1983, No. 5: p. 9-13.

Babaeva TA (1982) Use of the space photographs "Meteor" for desertification processes study in Turkmenistan. Probliemy osvoieniia pustyn, 1982, No. 6.

Centre National de la Recherche Scientifique (1977) World distribution of arid regions. Paris.

Centre National de la Recherche Scientifique (1976) World deserts and desertification. Paris.

Desert Institute (1993) Map of anthropogenic degradation of lands in Aral Sea basin scale1:2,500,000. Ashgabat.

Desert Institute (1985) Map of anthropogenic desertification of arid territories of USSR. Scale 1:2,500,000. Ashgabat.

Dregne HE (1977) Status of desertification in the hot arid regions. Compiled by FAO and UNESCO. - UN conference on desertification. Nairobi.

FAO (1981) Provisional methodolody for desertification assessment and mapping. Rome.
Kharin NG *et. al.* (1993) Criteria and methodology of assessment of desertification processes in the Sahel. Ylym, Ashgabat.
Kharin NG *et. al.* (1992) Methodical instructions on study of desertification processes of arid territories (on sample of Mongolia). Ashgabat.
UN (1977) Desertification map of the world. Nairobi.

9. The Technology of Desert Pasture Improvement

G.M. Mukhammedov, M.K. Durikov, N.T. Nechaeva

ABSTRACT: Sheep breeding is the most important branch of animal rearing in the Central Asian republics, and so it is of vital importance to control desertification by restoration of pastures. The Desert Institute through researches over many years has established sure ways of maximizing sustainable desert use. They work closely with farming authorities and farms. Detailed advice is given on how to cultivate pastures in desert soils including the optimal sowing ratio, types of seeds, sowing methods and phytoreclamation of pastures. The pastures are divided into various types: piedmont, sandy, clay, gypsum and artificial pastures.

The amount of non-irrigated lands in the Central Asian region is more than 90% of the total land. These are largely natural pastures used for grazing sheep and camels. They were formed and developed subject to the effect of the following factors: the presence of the established, natural, evolutionary type of shrub-grass and semishrub-grass pastures, a long, warm period in a year, and traditional experience of cattle breeding in arid zones.

The most important branch of animal breeding is Karakul sheep rearing. It provides considerable quantities of mutton, wool and, what is most important, karakul of a most beautiful pattern formed by differently shaped curls, richness of colors and tints of a hair coat. For the further development of Karakul farming, it is necessary to resolve a productivity problem and the quality of natural forage lands in the arid zone. This problem becomes increasingly important in the face of an accelerated growth of population, an increasing water deficit, and the necessity to enhance the role of cattle breeding in Central Asian countries as a raw material and foodstuff base.

The vegetation in a desert has a very important peculiarity: the capacity to renew and self-produce phytomass every year. Being an important component of the biosphere, it is responsible not only for forage resources of cattle breeding, but for the quality of human habitats in arid regions, a genetic diversity of flora and fauna. These natural multipurpose functions of vegetation of natural pastures can be realized only in conditions of rational management in ecologically admissible plans, and constant efforts to restore degraded pastures in order to improve their productivity.

Unfortunately, in the past two or three decades, the structure of vegetation has changed affected as it is by excessive anthropogenic and technogenic loads: highly productive shrub-grass pastures gradually turned into solely seasonal artemisia pastures, and these, in their turn, into hardly productive, grass pastures. In the vast Karakum and Kyzylkum deserts, sandy pastures around pits and settlements gradually become barkhan deserts.

It is necessary to take radical actions to ensure qualitative reconstruction of pastures in the arid zones. Nowadays both science and practice has a solid theoretical base and a technology sufficiently reliable for considerable enhancement of the productivity of arid pastures.

As a result of investigations over many years at the laboratory of pasture improvement the Desert Institute has developed efficient technologies to restore and improve pastures, to enhance their productivity, and has moreover outlined ways to maximize its rational use.

9.1 Pasture Types

Recommendations were prepared on the creation of the following types of long-term pastures:

- Pastures for autumnal-winter use. The necessity to establish autumnal-winter pastures is dictated by the fact that natural, ephemeral pastures are only of a seasonal use. They can be used mainly in spring and early summer. Having abundant forage resources in spring and summer, these pastures do not provide enough forage in autumn and especially in winter.
- Pastures for spring-summer use. It is recommended to establish these types of pastures in areas with the semishrub ephemeral and shrub grass vegetation. By the beginning of summer practically no feed remains on them. Under such conditions in order to cover a deficit in forage in spring and summer it is necessary to establish long-term pastures for spring-summer use.
- Year-round pastures should be created on all types of desert pastures where shrubs, semishrubs and grasses are growing, and that used in different seasons of a year.

It is advisable to create spring-summer and autumn-winter pastures on 15-20% of the territory of farms. Year-round pastures can be created on any area, because they have higher yields than natural pastures in all seasons of a year.

At present recommendations from the Desert Institute on intensification of pasture management are taken into consideration in perspective planning and day-to-day work of agricultural bodies, as well as sheep-breeding farms. An apposite example of this in Turkmenistan is the dekhan association Kalai-Mor of the Takhtabazar etrape, the association named after Magtymguly of the Gyaversky etrape; in Uzbekistan the state breeding farms "*Karnab*", "*Nishan*"; in Kazakstan – the state breeding farm "*Zadarinsky*". The introduction of scientific achievements facilitates the enhanced productivity and stability of the Karakul breeding.

The success of actions for the improvement of desert pastures is dependent to a great extent on properly and timely performed phytoreclamation works with regard to concrete ecological conditions. The principal objective of these works is to create favorable, water-physical conditions for the accumulation of moisture, nutritive substances, and weakening competition on the part of natural vegetation.

A system of agricultural practices aimed at the improvement of pastures includes the following: a choice of a site, methods of soil cultivation, sowing rates, methods and technique of sowing, depth and methods of seed embedding, and care after plantings.

A site for phytoreclamation works is chosen taking into consideration specific soil, hydrological, climatic and pasture-forage conditions. Highly productive pastures should be established, first of all, on areas with poor botanical composition of vegetation caused by overgrazing of cattle or some other anthropogenic factors, as well as on pastures which should be used in other seasons.

A correct and timely soil cultivation that ensures accumulation of the adequate amount of moisture in soil for a winter period is of great importance. Depending on the type of desert lands, the mechanical composition and physico-mechanical properties of ground soil, preparation for sowing includes ploughing, disking and harrowing.

In a Central Asian desert the main precondition for getting good seedlings is to sow plants in winter months (January-February) and embed seeds. Embedding creates the best conditions for seed germination, protects them from mechanical damage, prevents them from blowing away by wind and dragging away by insects

9.2
Technology of Pasture Improvement in Piedmont Deserts

Gigantic areas (more than 15 million ha) in Central Asia can be classified as a desert zone. Here spring-summer pastures with grass vegetation are mostly found. Excessive grazing in spring and summer creates a difficult situation concerning forage for Karakul breeding farms in autumn but more especially in winter.

Piedmont areas of Paropamiz (Badkhyz, Karabil) and Central Kopetdag feature a hilly relief with low ridges (their altitude over the sea level is 300-900 m, of some ridges 1200 m).The principal types of soils here are typically light, loamy-sand serozems underlain by sands and loamy sands. Sands are developed only around pits and were formed as a result of overgrazing that led to the destruction of a grass mat.

Climatic conditions of piedmont deserts in Turkmenistan are more favorable for phytoreclamation works in comparison to other natural regions in the desert zone. Precipitations vary from 150-250 mm a year. The average air temperature is 14°C, the average air temperature of the warmest month is 28°C, and that of the coldest month 1.7°C. An absolute maximum of the air temperature is 46°C. In spite of its southern location the absolute minimum of the air temperature is very low at minus 36°C.

Climatic conditions in the piedmont desert in Uzbekistan are characterized by the following indices: a dominating soil type i.e. light serozems and gray-brown coarse textured soils (loamy sands, light and medium loams), and low salinity. The average annual precipitations range from 130-160 mm. The average annual air temperature is 16°C. In June-July it is 40-45°C, while in January it goes down to minus 20-30°C.

Yields of pastures in piedmont deserts are 0.5-3 quin/ha of air-dry mass (1 quintal=100kg) . In order to enhance and stabilize productivity of these pastures, so as to reliably provide sheep with forage throughout a year, it is necessary to create enriched pastures of spring-summer and year-round use. (Plate 9.1.).

The technology of creating long-term, highly productive pastures includes the following.

Depending on the relief and attempting to avoid water and wind erosion, soils should be cultivated at right angles (90°) to prevailing wind directions and across slopes giving preference to valley-like hollows, gentle slopes and flat plateaus.

Cultivation of a dense, sedge-meadow grass mat 25-30 cm thick is conducive to better percolation of precipitations deep into the soil and to its slow evaporation. By mid-April the cultivated plot accumulates up to 80% of moisture of the total quantity of winter and spring precipitations, while virgin lands accumulate about 50% of moisture.

Cultivation quality depends on ploughing depth, moisture content, solidness of soil and applicable methods of cultivation. It is recommended to cultivate saline, low- and medium-saline, loamy sands and loams to a depth of 20-22 cm, and thin gypsum soils up to 15 cm, with subsoiling to 30 cm. For creation of winter pastures on lands with a non-projective layer, it is necessary to plough lands to a depth of 15 cm, because deeper cultivation can penetrate the whole non-projective layer of grass; the danger is that a low-nutritive horizon becomes exposed which is undesirable. Creating long-term pastures of spring-summer, autumn-winter and year-round use the chosen plot must be divided into strips. A width of ploughed strips and spacings among them depends on a life form of plants sown on strips. For sowing of shrubs and perennial grasses, a strip width should be 4.5 m and spacings 12 m. The cultivated area should take up 27% per ha. While sowing semi-shrubs the recommended width of strips is 6 m, and spacings among them 12 m. The ploughing area should be 33%.

For deserts of Central Asia the best time for ploughing is November-December, when precipitation starts falling. However, multiple investigations have shown that good results can be obtained by ploughing in February before sowing with simultaneous harrowing.

Under extreme desert conditions the timely sowing of seeds is conducive to getting sprouts of the desired thickness and to their further normal growth and development. Depending on climatic conditions in a year sowing should be done in winter or early spring, beginning from the third decade of January to the second decade of February. During long cold winters, sowing should be conducted in late February or even in early March.

For rational use of seeds and soil moisture it is necessary to define before sowing the optimal rate of seeding that will ensure the required thickness of a plant stand.

The optimal sowing rate can be estimated by the formula below:

$$K = \frac{H \cdot 100}{x}$$

K - the quantity of seeding material [kg/ha]
H - the recommended sowing rate assumed for 100% economic fitness [kg/ha]
x - economic fitness of seeds [%]; it can be estimated by the following equation:

$$x = \frac{A \cdot B}{100}$$

A - seed purity, %
B - laboratory germination of seeds, %.

Seeding material is usually a mix of good seeds with other parts of plants (leaves, immature fruit, small twigs), that is why in estimation of a sowing rate it is necessary to take into consideration a share of pure seeds in a mix that is expressed in percentage to the total weight.

Taking into consideration a field germination capacity of seeds and adaptability of plants in the first and subsequent years, as well as knowing an absolute weight of 1,000 seeds, one can estimate the sowing rate assuming 100% fitness of seeds. Consequently, knowing these elements one can estimate the full sowing rate by a formula.

Table 9.1. gives sowing rates of seeds of most promising varieties of forage plants recommended for cultivation in piedmont areas.

Seeds of forage plants that are to be sown should conform in their sowing qualities to technical requirements. One should sow seeds only of the first and second classes. While creating perennial pastures using different varieties and life forms of plants, the sowing rate of seeds should be calculated using respective rates for single-component sowings in proportion to a percentage ratio of these varieties in a mix.

Correlation of varieties depends on a type of created pasture:

- perennial pastures for spring/summer use: semishrubs 75%, grasses 25%;
- perennial pastures for autumn-winter use: shrubs 30%, semishrubs 70%;
- erennial pastures of year-round use: shrubs 20%, semishrubs 60%, grasses 20%.

There are different methods of plant sowing on desert pastures: overall, random, row, wide row, check-row and others. When sowing pastures it is better to apply overall sowing, when creating haylands and seed plots it is recommended to apply a row sowing that makes it possible further on to use agricultural machinery.

Table 9.1. Sowing rates of plants (kg/ha)

Plants	Cultivation		
	overall (100%)	by strips (50%)	by strips (33%)
Shrubs			
Calligonum rubens	8 kg/ha	4.0 kg/ha	2.7 kg/ha
Haloxylon persicum	7 kg/ha	3.5 kg/ha	2.4 kg/ha
Haloxylon aphyllum	7 kg/ha	3.5 kg/ha	2.4 kg/ha
Salsola paletzkiana	9 kg/ha	4.5 kg/ha	3.0 kg/ha
Salsola richteri	9 kg/ha	4.5 kg/ha	3.0 kg/ha
Halothamnus turcomanica	9 kg/ha	4.5 kg/ha	3.0 kg/ha
Ephedra strobilaceae	6 kg/ha	3.0 kg/ha	2.0 kg/ha
Semishrubs			
Kochia prostrata	7 kg/ha	3.5 kg/ha	2.4 kg/ha
Geratoides papposa	9 kg/ha	4.5 kg/ha	3.0 kg/ha
Salsola orientalis	7 kg/ha	3.5 kg/ha	2.4 kg/ha
Artemisia badhysi	8 kg/ha	4.0 kg/ha	2.7 kg/ha
Artemisia turanica	8 kg/ha	4.0 kg/ha	2.7 kg/ha
Perennial grasses			
Agropyron sibiricum	10 kg/ha	5.0 kg/ha	3.3 kg/ha
Astragalus agameticus	9 kg/ha	4.5 kg/ha	3.0 kg/ha
Astragalus maximowiczii	10 kg/ha	5.0 kg/ha	3.3 kg/ha

Seeds can be sown by machines or manually if the relief prevents from application of special sowing machines. Seeders SST-3, SZT-3.6, SUT-47 and IPS-0.4 can be used for mechanized sowing. When seeding machines are used it is recommended to apply a wide-row method.

A depth of seed embedding in soil. Field germination of seeds depends to a great extent on a depth and method of seed embedding in soil. The most effective practice that ensures a 2-5 times better field germination of seeds is after-sowing packing, especially in combination with after-sowing harrowing with packing.

In piedmont deserts the optimal depth of seed embedding is 0.5-1 cm for *Haloxylon aphyllum, H. persicum, Kochia prostrata, Eurotia ceratoides*; 1-1.5 cm for *Salsola orientalis*; 1.5-2 cm for *Salsola paltezkiana, S. richteri, Ephedra strobilaceae*, different species of *Calligonum* and *Astragalus*.

9.3 Technology of Pasture Improvement in Sandy Deserts

Pastures in sandy deserts of Central Asia cover a greater part of the territory. Their total area is 44.8 ha, of which 26 million ha are in Turkmenistan.

A specific feature of a relief of sandy deserts is the presence of high ridges stretching as a rule from the north to the south. Their height varies from 3-25 m, while a width of inter-ridge depressions reaches 1.5-2 km. The nature of inter-ridge depressions is different. Quite often depressions are overlain by smaller transverse eolian ridges, thus, mainly the relief is more hillocky and honeycomb-like. Depressions are often represented by takyr sand ridges and solonchaks.

Climate in the sandy desert Karakum is sharply continental, the average annual air temperature here being 14.8-16.4°C. The absolute highest air temperature (up to 50°C) is observed in July-August, and the lowest (-20°C to -23°C) in January-February. The annual precipitations in its different parts is 124-148 mm. They fall mostly in winter and spring. Air aridity is so high that already in March the relative air humidity is as low as 47%. On some very hot days when the air temperatures reaches 40-47°C, the air humidity can drop to 5%.

In sandy deserts most widespread are desert sandy soils. They are composed of medium- and finely-grained nonsaline sands. A specific feature of these soils is availability of a compact densely-rooted horizon overlain by a thin loose sandy layer. Natural fertility is low, the humus content is up to 0.5%.

Vegetation of a sandy desert is rather diverse and are represented by different life forms. Most widespread are shrubs and small grasses (ephemerals and ephemeroids) vegetating in winter and spring. Semishrubs (*Artemisia spec.*, *Salsola spec.*), loose-shrub cereals and annual plants that vegetate in summer and autumn are of limited distribution here, but even so they are important.

Forage reserves of sandy pastures are low and their capacity is insignificant. The lowest forage reserves are in the Northern Karakum, somewhat higher in the Central and South-Eastern Karakum. On different pastures an average annual eatable reserve varies from 0.6-2.3 quin/ha.

In a sandy desert, depending on the specific geomorphological conditions, life forms of plants and fixation by vegetation, the appropriate methods of surface and radical improvement of pastures should be applied.

Surface improvement of pastures is accomplished without soil cultivation by way of undersowing of forage plants with various life forms in originally thinned grass stand. Seeds are embedded by driving a sheep flock over a pasture.

On a sandy surface where soil is carried away by wind and is poor in nutritive substances and fine earth, it would be most efficient to sow palletized seeds. Heavy crumbs-pellets are obtained by putting seeds into a thick solution of sand and clay (manure), then the seeds are dried. Pellets are scattered manually because barkhan sands are impassable for machinery. Palletized seeds are not easily blown away while clay and manure entering them ensure feeding for seedlings in the first weeks of their life.

Good results are obtained from planting of seedlings of *Haloxylon persicum*, *Salsola paletzkiana*, and cuttings of *Calligonum spec.*. (Plate 9.2.)

Radical improvement of sandy pastures is advisable, first of all, on areas with a flat relief where machines can be used. Territories with poor vegetation, degraded as a result of unwise use, sands near pits fixed with grass vegetation, but devoid of shrubs and semishrubs should be improved first.

Soil cultivation should go in with harrowing or packing with sprocket packers, thus, ensuring leveling of the cultivated surface, better accumulation and use of soil moisture.

In order to prevent wind erosion the sandy soils should be cultivated by strips running perpendicularly to directions of prevailing winds. Soil is cultivated for sowing of plants by disking to a depth of 8-12 cm, or ploughing to a depth of 16-20 cm; that will do away with sedge and ensure adaptability of seedlings.

In the conditions prevailing in the Karakum, on small hillocky sands fixed with grass vegetation (especially *Carex physodes*) the width of cultivated strips should be 3-6 m with interspaces of virgin lands 15-30 m wide i.e. it is advisable to use a ratio of ploughlands and virgin lands as 1:5.

In sandy deserts the best sowing period is January-February, as most often mixes of various plants are sown. In regions with variable winter sowing can be done in early spring (late February - early March) after moistening sands to a depth of 10-15 cm.

The most suitable sowing rates that will ensure seedlings with an adequate stand are the following: *Haloxylon aphyllum* (5 kg/ha); *H. persicum* (6 kg/ha); *Salsola paletzkiana* and *S. richteri* (12 kg/ha); *S. orientalis* (6 kg/ha); *Kochia prostrata* (4 kg/ha); *Artemisia kemrudica* (0.1-1 kg/ha); *Calligogonum spec.* (15 kg/ha); *Halothamnus turcomanica*, *Ephedra strobilaceae* and *Astragalus maximowiczii* (8 kg/ha); grass plants (2-3 kg/ha) assuming 100% economic suitability of seeds. (Plate 9.3.).

Embedding of seeds to an optimal depth creates better conditions for germination, protects them from being carried away with the wind, or from dragging away by insects and birds. It should be done by a harrow with simultaneous packing. In cases when it is impossible to use machines the seeds should be embedded by driving a flock of sheep. An embedding depth of small seeds is 2 cm, and of larger seeds 3-4 cm.

Choice of a sowing method is dictated by a relief. In particular, for a broken terrain of a sandy desert a manual sowing by throwing seeds under a harrow from a cart or cross-country trucks is most matching and efficient.

9.4
Technology of Pasture Improvement in Clay Deserts

Pastures in clay deserts of Central Asia cover more than 20 million ha, of which in Kazakstan 13.5 million, in Uzbekistan 2.6 million, in Turkmenistan 4.5 million. Great areas of clay deserts are found in sandy deserts of the Karakum and Kyzylkum, in ancient deltas of the Amudarya, Syrdarya, Myrghab, Tedjen, Kashkadaria rivers. They are also met in combination with other types of deserts and are of spot distribution.

A relief of the greater part of a clay desert is distinguished by its flatness. Quite often this is an absolutely flat surface seldom cut by shallow beds of interim surface streams with a very insignificant slope.

A soil cover of a clay desert is represented by mostly takye-like and more seldom light serozems composed of loess-like loams and sometimes loamy sands. Takyr sand ridge soils are developed on ancient alluvial and proluvial plains. In their texture these soils are usually clay, heavy-loamy or medium-loamy. In view of a compact surface the wetting of these soils by precipitation is poor, water often stagnates on the surface and is lost for evaporation.

Takyr sand ridge soils are poor in nutritive substances, the humus content is 0.3-0.8%, water soluble salts 2-2.5%. In takyr sand ridge soils formed as a result of sanding and destruction of takyr sand ridges a humus content in takyr sand ridge-like soils can be as low as 1%. On such soils semishrub-grass vegetation suitable for grazing sheep and camels only in autumn and winter is growing. An average annual reserve of edible feed on semishrub pastures varies from 0.9-1.7 quin/ha. However, considerable territories in a clay desert are devoid of a shrub vegetation, and typical takyr sand ridges do not have any higher vegetation on which cattle is feeding.

Phytoreclamation measures together with special agricultural practices can be conducive to turning takyr sand ridges and takyr sand ridge-like terrains into highly productive pastures. When creating such pastures, one should distinguish between typical takyr sand ridges devoid of any higher vegetation where a surface runoff of precipitations is forming, and takyr sand ridge-like terrains slightly sanded on a surface where a surface runoff is not formed. On typical takyr sand ridges the phytoreclamation measures should be aimed at creation of artificial pastures applying water-accumulating furrows (Photo 9.1.) and cultivation of *Haloxylon aphyllum, Salsola orientalis, S. gemmascens* and other valuable forage crops adapted to growing on heavily saline soils.

On takyr sand ridges, takyr sand ridge-like and solonetz gray-brown and gypsum soils heavily compacted on the surface, it is necessary to cut water accumulation furrows to a depth of 35-40 cm by a single-moldboard reclamation plow PPN-40 or PPN-50. Furrows should be cut in autumn so that water from snow thawing or precipitation runoff can accumulate in them, or in early spring when soils are still wet. They are spaced 10-15 m so that rainfall runoff from virgin land interspaces ensured adequate moistening of soil under furrows.

In the Karakum and Kyzylkum, the areas of clay deserts can be found amongst large sandy terrains that are sources of sand blown by wind over a clay desert. Sand accumulation furrows are cut in the same way as water accumulation ones applying a reclamation plow. Depending on intensity of a wind-sand flow the furrows are in every 5, 10 and 15 m. Sand accumulating furrows should be cut in spring awaiting their more intensive filling with sand in a summertime.

Advisability of sowing by furrows is connected with the favorable texture of takyr sand ridges in the Karakum, because at a depth of 50 cm loamy sandy horizons appear, underlain by sandy horizons facilitating quick seepage of precipitation from a furrow.

It is recommended to develop small takyr sand ridges of no watershed significance for pastures. At first plowed-moldboard furrows should be cut by a three-body plow mounted on a tractor DT-54. Furrow width at a virgin land level is 50 cm, depth 30-35 cm, width of moldboard plowing is 100 cm. In general, width of a furrow and moldboard ploughing is 150 cm, a spacing among furrows is 10 m. Here 80% of the total takyr sand ridge area remains intact. Precipitation

Photo 9.1. Cutting of water accumulation furrows on takyr

runoff from inter-barkhan areas gets into water accumulation furrows via water outlets (Fig. 9.1.).

In order to preserve large takyr sand ridges as watershed areas for phytoreclamation, it is recommended to use only marginal sites that have better reclamation conditions. In this case on takyr sand ridges lying lower than the main slope, they cut the main water distribution furrow and double-moldboard furrows will be cut from it spaced 10-15 m from each other, after which seeds of forage crops are sown into them.

The purpose of development of sanded takyr sand ridge-like plains is to enhance their productivity by undersowing to the available poor vegetation cover, the missing tree-shrub plants. As there is no precipitation runoff, seeds should be sown into ploughed strips. Ploughing depth should be 30-35 cm, width of ploughed strips 4-6 m, and spacing between them 15-20 m. The best period for sowing in a clay desert is winter - December-January.

To attain the desired stand of plants under severe soil conditions of takyr sand ridges and takyr sand ridge-like surfaces it is recommended to apply the following sowing rates (assuming 100% economic fitness of seeds): *Haloxylon aphyllum* (7 kg/ha); *Salsola paletzkiana* and *S. richteri* (12 kg/ha); *Halothamnus turcomanica* (10 kg/ha); *Artemisia kemrudica* (1 kg/ha); *Salsola orientalis* (7 kg/ha);

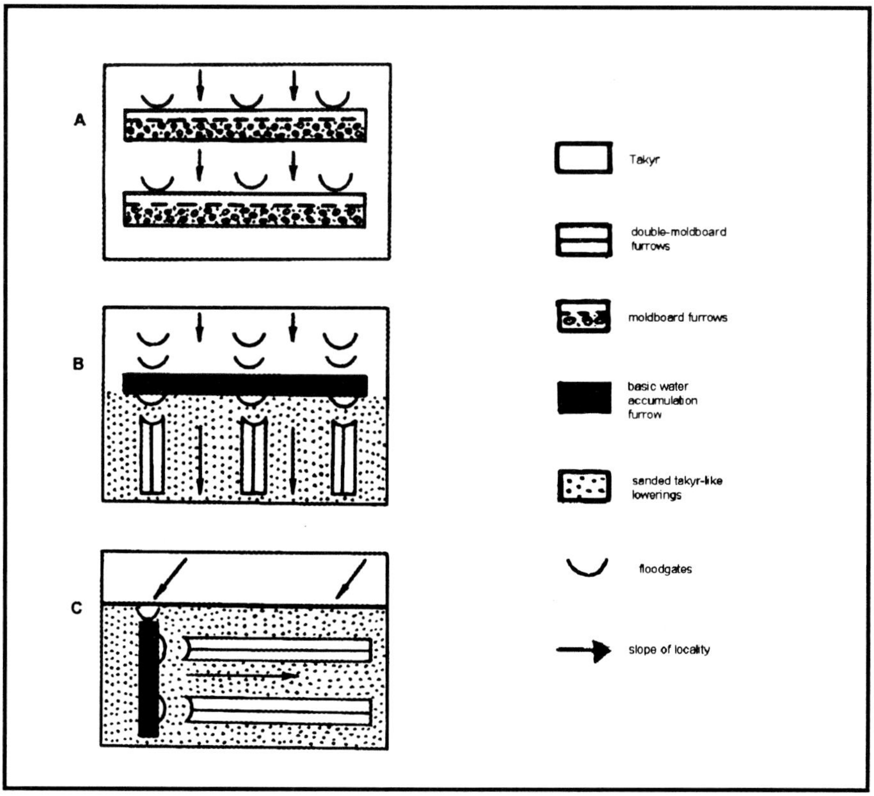

Fig. 9.1. Variants of cutting furrows on takyr: (**A**) cutting single-moldboard (**B**) and (**C**) variants of cutting double-moldboard furrows

Salsola gemmascens (5 kg/ha); *S. sclerantha* (4 kg/ha). A mix of seeds should be sown in the following ratio: shrubs and semishrubs 70%, grasses and annual thistle 30%.

Seeds should be embedded into soil by a light harrow with simultaneous packing. The optimal embedding depth for large seeds is 1.5-2 cm, for small seeds 0.5-1 cm.

9.5 Technology of Pasture Improvement in Gypsum Deserts

Pastures of gypsum deserts are widespread on the Ustyurt, in Zaunguzje and South-Western Kyzylkum. Their total area in Central Asia reaches 38 million ha, of which about 5 million ha in Turkmenistan.

The soil hydrological conditions and a relief of a gypsum desert are characterized by great diversity: hillocky and loose ridge sands, drainless solonchak hollows, residual uplands, sand ridges, and takyr sand ridges). Diversity of relief forms is conducive to uneven distribution of precipitation, and wetting of a soil cover affects the vegetation development.

A surface of gray-brown gypsum soils is often heavily covered with crushed stone and small pebble. A thickness of a gypsum layer is in some places 40-50 cm from the surface, and often even more 60-115 cm. Gypsum soils contain carbonates and sulfates in great amounts. A humus content is insignificant 0.3-0.5%. Depending on meteorological conditions and terrain relief, precipitations can wet soil to a depth 40-80 cm.

In a gypsum desert a vegetation cover is represented by different varieties: ephemeral-ephemeroid, semishrub-grass, and shrub-grass communities. Low productive pastures of seasonal use are to be improved. For this purpose pasture shelterbelts 3-5 m wide alternating with natural vegetation belts 30-50 m wide should be made.

Inhomogenity of a soil and vegetation covers in a gypsum desert makes it necessary to apply different soil cultivation practices. The following methods are recommended: traditional moldboard plowing to a depth of 20-22 cm; deep plowing to a depth 30-35 cm; cutting of sand accumulating furrows. On light-textured soils with ephemeral and ephemeral-*Salsola sp*-*Artemisia sp.* vegetation a presowing soil cultivation should be made to a depth of 20-22 cm with subsequent harrowing.

On gypsum gray-brown soils with unfavorable physico-mechanical properties a shallow plowing leads to the movement up to the surface of clayey particles that are swelling. The water permeability properties of soil become worse, precipitation evaporates quickly, a compact crust forms on the soil surface, seeds do not germinate properly. Under such very unfavorable conditions it is recommended to apply deep plowing to a depth of 30-35 cm.

Sand accumulating furrows should locate perpendicular to a direction of prevailing winds. They are cut by a reclamation plow PPN-40 and PPN-50 to a depth of 30-35 cm, 50 cm wide, in every 5, 10 and 15 m. Sand facilitates improvement of water-physical properties of soil. Apart from sown seeds of shrubs and semishrubs, seeds of ephemerals get into the same furrows, thus, the vegetation cover becomes more diversified.

The following sowing rates are recommended: *Kochia prostrata* (4 kg/ha); *Halothamnus turcomanica* (10 kg/ha); *Camphorosma lessingii* (5 kg/ha); *Salsola orientalis* (8 kg/ha); *Haloxylon aphyllum* (6 kg/ha); *Ceratoides papposa* (10 kg/ha); *Artemisia diffusa* (1 kg/ha).

Seeds are sown in the following proportions: shrubs 20-25%, semishrubs 50-60%, grasses and annual thistle 25-30%.

9.6 Care of Artificial Pastures

Artificial pastures do not need additional soil treatment but proper care of them – like after natural pastures – include their rational management.

It is important to use artificial pastures in due time. If they comprise early-maturing quickly growing plants (*Halothamnus, Artemisia, Kochia*), then grazing on them should be started in the second or third year, and if they comprise slow developing varieties (*Haloxylon, Ephedra*), then in the third or fourth year when plants get stronger. Biting off of young sprouts of shrubs and semishrubs at a moderate load ensures formation of low shrubs accessible for sheep. Such branchy, short shrubs provide much more forage and are more convenient for cattle.

In the first years it is advisable to start grazing in winters, and in subsequent years with regard to seasonal suitability of created pastures in the seasons when the necessity arises. Artificial pastures can be grazed annually with a moderate load, because eating young twigs in moderate quantities cannot cause damage.

Investigations conducted for three years into the effect of sheep grazing on some types of shrubs and semishrubs, have indicated that biting of sprouts of the first year in winter to a length exceeding their natural withering length is conducive to increment of a forage yield in the next year. On the remaining parts of sprouts, earlier dormant buds become alive and in subsequent years they form young sprouts in a greater quantity than in a control treatment.

An optimal load of animals on pastures is determined by estimations of forage productivity of each pasture plot and requirements of grazed cattle in nutritive substances. Under such conditions the rational management of pastures is attained by using in practice the optimal norms of cattle load.

A moderate use of the vegetation cover (feeding 65% of the yield to animals annually) is one of the ways to maintain productivity of pastures and a favorable ecological situation for many years ahead. Under moderate grazing, a typical composition of the vegetation and its yields remain intact.

Under moderate grazing, density of a grass stand diminishes, while good trampling of seeds into soil will ensure good renovation of plants. Therefore, moderate grazing has a positive effect on artificial pastures.

Shrub thickets should not be cut out and stubbed because this leads to their quick and complete destruction. If artificial pastures are not grazed at a young age, and they have had enough time to turn into high plantings, then it is necessary to rejuvenate them by cutting branches at a height of one meter from the soil surface. Cutting should be done in winter (December, January) which is conducive to springtime rewakening of dormant buds in the lower part of shrubs and formation of many young branches. Therefore, low shrubs form a "pasture shape". The cut-off branches can be used as firewood – which is in great deficit here. Knowing of the high demand for wood, it is necessary to protect plantings from grazing within 6-7 years, and then cut branches for fuel, while newly grown sprouts are used for feeding. Moderate grazing and cutting of branches off the shrubs serve to rejuvenate them and increase the lifetime of plants.

One of the most important conditions for pasture improvement is provision of an adequate plant stand in strips. Excessively dense stand, say, of *Haloxylon* will lead in 4 or 5 years to closing of canopies, to complete shadowing and disappearance of ephemerals and thistle under the canopy of *Holoxylon* plantings. Therefore, dense *Holoxylon* plantings in the second or third year should be thinned ensuring their even distribution.

In the course of long use of sown pastures (for 12-15 years), natural thinning out of plantings is observed and their yields become 60-65% lower in relation to the original. In this context it is necessary to rehabilitate them. In November-December the soil should be cultivated with disk harrows LD-5 or LD-10 in two tracks with subsequent undersowing of seeds proceeding from a rate 35-40% of the original rate. Rehabilitated plots can be used for grazing from autumn of the next year.

Improvement of pastures is one link in a chain of action aimed at intensification of desert cattle breeding. Efficiency of phytoreclamation works can be seen not only in the increased yields of forage crops to 12-20 quin/ha, but in additional earnings from sheep breeding. Net profit obtained per 100 ha of improved pastures is 3-5 times higher than per 100 ha of natural pastures. A repayment period of investments into creation of long-term pastures is 4 or 5 years. During two or three decades the improved pastures maintain high productivity and ensure additional profit.

10. The Agrometeorological Foundations for Prediction of Pasture Yields

M.N. Nurberdiev

ABSTRACT: To maximize pasture yields for the most important desert animals, accurate meteorological predictions for up to six months in advance are necessary. The Desert Institute has tested various methods and opted for those described. Essential instruments of calculation are the radiation aridity index, the moisture supply index and wetting coefficients each of which is elucidated. Cloudiness, adaptability of seedlings, and the accumulated storage of accessible moisture in soil also need to be mathematically evaluated. For each of these factories, examples are provided using typical figures from the Central Asian region. Autumnal cloudiness of the current year and the adaptability of seedlings are valid predictive indexes for the following years pasture yield, and so in this way farmers can be accurately advised.

Increasing cattle breeding productivity and cattle stock in Central Asia is possible only under efficient use of natural pastures. To attain this it is necessary to build up in-depth knowledge of natural climatic factors influencing all aspects of animal husbandry: pasture forage reserves, grazing terms, phytoreclamation and others.

In dry years the yield of pasture grasses in Central Asian deserts can drop to 30-50 kg/ha, while in favorable years it can be as high as 400-500 kg/ha. In some unfavorable years the animals need to be fed in their quarters. In cattle farms where no serious actions were taken in critical years to organize such feeding of animals, in winter the animals not only suffered from strong depletion, but high mortality of sheep was observed. Therefore, climate considerations should be taken into account and rationally applied in agricultural production.

Productivity of the natural vegetation cover (Fig. 10.1.) is closely connected with the heat and water balance on the surface. Consequently, a zonal productivity of live organic matter depends on heat and water correlation. In this context many agrometeorologists in Turkmenistan, Uzbekistan and Kazakstan while developing forecast methods used as a basis the principal climatic factors (water and heat supply) during a vegetation period of pasture plants. (Fedoseev 1964, Gringof 1965, Arbykov 1968, Nurberdiev 1976). As a result the Desert Institute of Turkmenistan created and tested different methods of prediction of agrometeorological conditions on which the yield of pasture plants is dependent. Forecasts were made for 1.5-2 months at large. However, today cattle breeders need a forecast that can

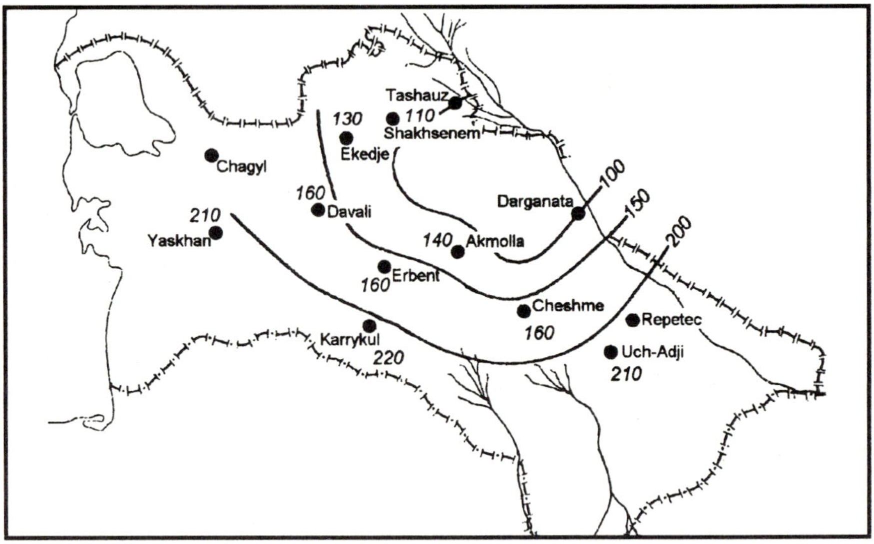

Fig. 10.1. Isoline of average spring grass crops, kg/ha

predict the expected yield of pastures for six months ahead. Our proposal is outlined in the following text.

10.1 The Radiation-Aridity Index

Central Asian deserts feature the highest disproportion between the thermal energy base (5,500°C) and a soil moisture content (100-148 mm). The radiation-aridity index (R/L) reflecting a ratio of heat quantity to moisture quantity reaches 6-7. Here a quantity of heat, R, is expressed by the amount of an annual radiation balance of surface activity. A moisture content, L, denotes calories necessary for evaporation of annual precipitation.

Such deviations of a heat-moisture correlation seriously affect plant growth. However, in spring a disproportion between heat and moisture alleviates because of low temperatures and concentration of precipitation and low evaporation. For instance, in different regions of the Karakum desert, in spring the radiation aridity index drops to 2-4, which creates favorable conditions for vegetation of pasture plants.

Therefore, the amount of precipitation for a wet period of a year (December through April) and air temperatures in spring months are principal indices that control the productivity level of pasture vegetation in different years.

Many agrometeorologists used quantitative appraisals of moisture and heat, correlation of these factors and intensity of their manifestations in different time periods as a basis for development of relationships. The scientists revealed that the aggregate demand of cultural plants in water is close to an evaporation figure, or is characterized by its steady ratio to a sum of deficits of air humidity. Water used for total evaporation (transpiration and evaporation form the soil surface) during a vegetation period of different agricultural crops varied in the Centre of Europe from 0.66 of a sum of mean daily deficits of air humidity.

10.2 The Moisture Supply Index

Estimations made for desert territories of Kazakstan have shown that during a vegetation period up till flowering, the relative demand in moisture with ephemeral plants in high-yield years was equal to 0.46 of the sum of deficits of air humidity (d), while for a period from flowering to ripening of seeds it was 0.22. Data for the Central Karakum coincides with indices for a desert zone in Kazakhstan. Therefore, a moisture supply index (PV) over ten days was estimated as the difference between the moisture storage in a dead layer of soil for a previous (V_1) and subsequent (V_2) ten day period plus precipitation (Os) for this period related to a half sum of average daily deficits of air humidity in mm (0.5 Σd):

$$PV = \frac{V_1 - V_2 + Os}{0.5\sum d}$$

An absolute value of the moisture supply index, PV, being 1.0 means that a grass stand is provided in accordance with its relative demand; PV > 1.0 means partial moisture supply of grasses. When PV equals 0.2 the yield increment in the Central Karakum is 10 kg/ha for a decade, PV=0.4 corresponds to 40 kg/ha, while PV=0.6-80 kg/ha.

However, until recently there has been no reliable data on soil wetting for many regions of a sandy desert. This prevents one from getting indices of moisture supply of plants. Therefore, the estimations of indices of moisture supply of plants used the amount of precipitation.

10.3 The Wetting Coefficient

A wetting coefficient (K) of sandy deserts e.g. the Karakum, is estimated as a ratio of the amount of precipitation (S_o) to a sum of mean, daily deficits of air humidity (Σd) for a wet period of a year. The beginning of a wet period is assumed to be the first ten days of autumn or winter when the amount of precipitation is higher than the evaporated moisture, and soil moisture content starts increasing. The end of a wet period is fixed by a time of maximum accumulation of dry, grass mass.

A yield of spring grasses is directly connected with the values of the wetting coefficient. The correlation coefficient $r = 0.823 \pm 0.039$ and a regression equation takes the form Y = 10.48 K-0.09, where Y is the highest yield of spring grasses, quin/ha; K is the wetting coefficient of a territory. An average K is equal to 0.18 with variations from year to year from 0.06-0.6. These variations are dependent on changes of precipitation in a wet period of a year.

Average many-year precipitation are decreasing from south to north from 140 to 120 mm. Nearly over the whole territory of the Karakum 74-82% of the annual rate of precipitation fall in winter and spring (December through April). A relationship between the highest yield of spring grasses and the amount of precipitation for the period December through April was defined earlier, and is characterized by the correlation coefficient r = 0.738-0.054.

The principal yield of the grass vegetation in the Karakum deserts is formed in spring i.e. most often in February-April. Precipitation in December through April differ greatly from year to year.

Recurrence of different levels of precipitation and yields of grasses in the Karakum shows vividly that recurrence of the amount of precipitation for a period December through April being less than 50 mm is 22%, or twice in ten years. The same is the recurrence trend of precipitation higher than 11 mm for the same period. In most cases (57% of years or in six years out of ten) there was registered precipitation that came close to the year average (over many years) figure (80 mm) deviating to plus and minus sides from the average. In this case precipitation whose level is 80 ± 30 mm, or within 51-110 mm are recurring six years out of ten.

Returning of years with different levels of grass yields is close to recurrence of different levels of precipitation. Low-yield years (less than 1.2 quin/ha) occur every three years, an average-yield every 5 years and high-yields every two years out of ten. The proportion of different grass yields increases disproportionately less to an increase in precipitation.

10.4 Cloudiness

One of the major tasks in development of long-term, agrometeorological forecasts of spring grass yields is the determination of the expected level of heat and moisture supply in a winter-spring period on which plant productivity depends. One of the critical factors influencing formation of a radiation and heat regime as well as moisture supply of deserts is cloudiness.

Solar radiation, the amount of precipitation and, consequently, soil and air temperature, humidity of soil and air, and then heat and moisture supply of plants; all these are connected with cloudiness. Therefore, the basis of the long-term, agrometeorological forecast in this paper is made by a sum of points of total cloudiness as a meteorological index that most comprehensively reflects maximum yields of pasture vegetation and information from a network of meteorological stations of the Hydrometeoservice.

A sum of points of total cloudiness can be replaced by a number of cloudy days in September and October. In order to get this value it is necessary to deduct the number of bright days by total cloudiness from all days in October (31) for the Eastern and Central Karakum and from all days in September (30) for the Northern Karakum (Fig. 10.2.).

The correlation coefficient of these relationships is as follows,

$$r = 0.70 - 0.72 \pm 0.07 - 0.10,$$

while a regression equation will be, respectively,

$Y = 0.07x + 1.0$ for the Eastern Karakum
$Y = 0.14x - 0.04$ for the Central Karakum
$Y = 0.10x + 0.26$ for the Northern Karakum

where x is a number of cloudy days in October.

Applying a new method, a forecast was prepared in early November of 1998 for the yield of the next year. Study of the forecast reality using the available observations made over many years has revealed that a deviation of the actual yield from that forecasted is on the average 0.3-0.5 quin/ha (1 quintal=100 kg).

Successful implementation of phytoreclamation works makes it necessary to single out from complexes of major natural factors those which control vegetation of plants sown in a given area. Climatic factors are critical among all natural factors, and include heat and moisture supply of a territory: these are especially important for development of different varieties of plants. In the Karakum heat supply of the territory is high enough, but a moisture deficit is very acute. That is why moisture supply of the Karakum territory determines the possibilities for implementation of phytoreclamation works in different years.

10.5 Adaptability of Seedlings

Different amounts of precipitation by years create unfavorable conditions for adaptability of seedlings, thus, first of all, it is necessary to estimate recurrence of favorable and unfavorable years. Recurrence of good adaptability of seedlings on small, hillocky sands is 3 years, average adaptability is 4 years, and unsatisfactory 3 years out of 10.

Depending on the relief forms of a terrain the recurrence of different (in terms of moisture supply) years varies, so each relief element should be considered separately.

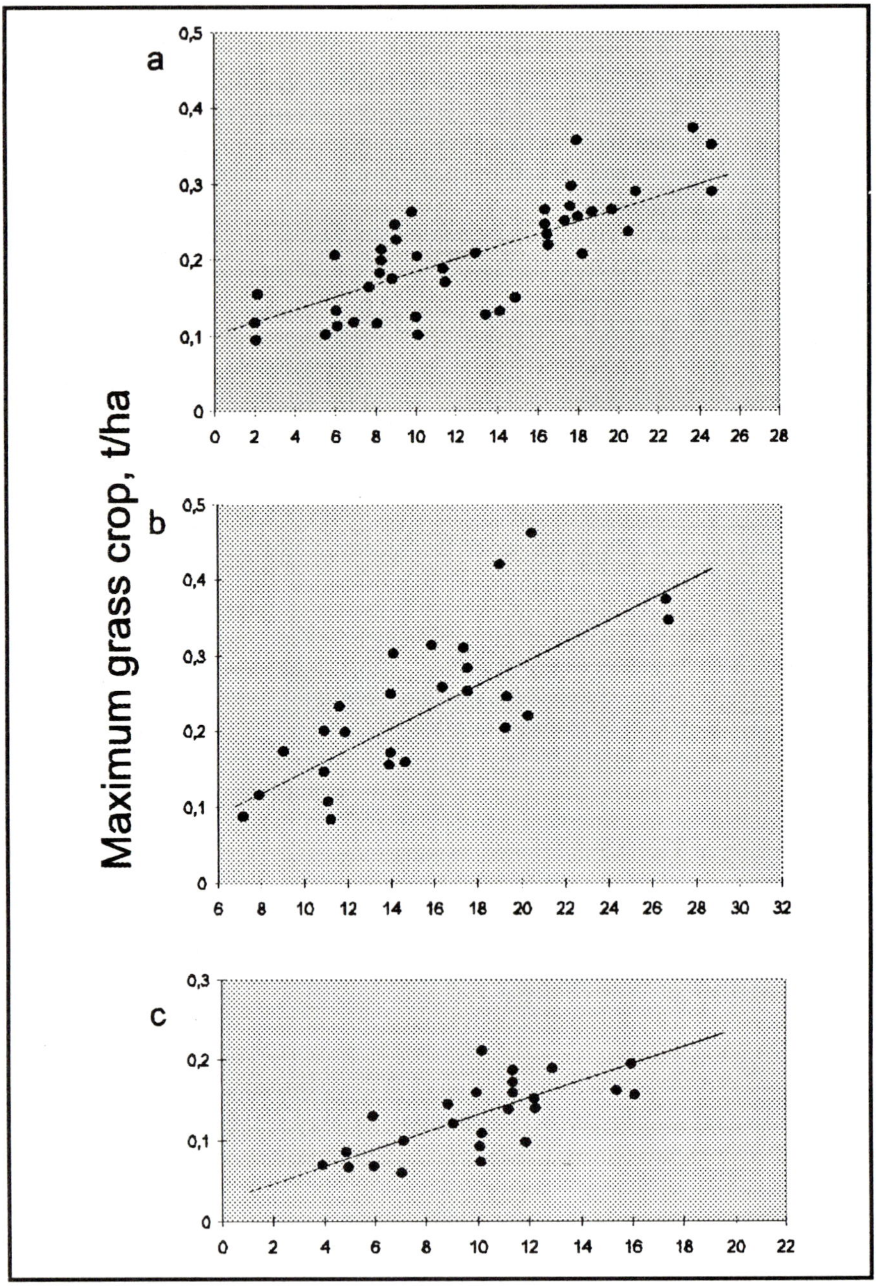

Fig. 10.2. Connection of spring grass crop with the number of clouded days in Eastern and Central Karakum (**a**, **b**) and September - in Northern Karakum (**c**)

Table 10.1. Deviations in DSW and accessible moisture storage within a one-meter soil layer (early spring)

Relief element	Accessible moisture storage within 1-meter layer (Y_1)	DSW (Y_2)
A contact zone of the takyr-like depression with sands	Y + 12 mm	Y – 25 cm
Small hillocky sands	Y – 0 mm	Y – 0 cm
Eastern and western slopes of a sand ridge	Y – 14 mm	Y – 16 cm
Unfixed top of a sand ridge	Y – 8 mm	Y – 10 cm
Sanded takyr-like depression	Y + 12 mm	Y – 25 cm

Taking into consideration sharp variations of precipitation within a year and by seasons, it is necessary to develop methods of long-term, agroreclamation forecasts of expected adaptability of seedlings. The principal agroreclamation index that controls adaptability of seedlings is a depth of soil wetting (DSW) or an accessible water storage within a one-meter layer of soil during a sowing period (early spring).

An accessible moisture storage in a one-meter layer of soil and DSW in early spring are mostly dependent on the number of cloudy days in October. Correlation coefficients are r = 0.70-0.80, while the corresponding regression equations become as follows,

$$Y_1 = 3.9x - 30$$
$$Y_2 = 5.9x - 53,$$

where, Y_1 is an accessible moisture storage within a 1-meter layer in early spring [mm] and Y_2 is the DSW in early spring.

Having substituted the values of x into respective equations in early November, we can estimate the awaited DSW, or a soil moisture storage in early spring of the next year. The awaited DSW or a soil moisture storage are estimated by equation for a reference plot of low, hillocky sands. For other elements of a relief, deviations of these indices are given in the Table 10.1.

DSW values or a soil moisture storage, estimated by equations for small hillocky sands are corrected for respective elements of a relief. As a result they obtain DSW or a soil moisture storage expected in early spring in the next year.

During the first years, the overground and subsurface organs of desert plants develop in the lower terrestrial layer of air, and upper layers of soil which have the highest daily, monthly and seasonal values of meteorological elements. During the first year the plant should overcome a critical "barrier" of agrometeorological conditions that control their adaptability and growth.

The critical factor for phytoreclamation works in a desert is the verification of optimal terms for sowing. The most rational sowing time in the Karakum is early spring (a mean daily temperature of air is steadily above +5), because the conditions of soil wetting in the autumn-winter period (up to early spring) are conducive to successful, phytoreclamation works.

Table 10.2. Agrometeorological indices for various relief elements in the Central Karakum and their assessment for adaptability of seedlings (in the presowing period from Nov. to early spring, when air temperature is higher then 5°C)

Relief element	Adaptability of seedlings	Precipitation Nov.-Febr. [mm]		Y_1		DSW during sowingperiod [cm]	
		Average	Varying	Average	Varying	Average	Varying
Contact zones of takyrs and sands	good		60		50		40
	average	46	35-59	50	30-49	50	26-39
	unsatis.		34		29		24
Inter-ridge takyr hollows	good		60		75		50
	average	46	45-59	60	50-74	50	35-49
	unsatis.		44		49		34
Inter-ridge small hillocky sands and lower fringes of eastern and western slopes of a sand ridge	good		60		50		60
	average	44	45-59	38	35-49	75	45-59
	unsatis.		44		34		44
Middle of western and western slopes of a sand ridge	good		90		70		90
	average	44	55-89	24	55-69	60	75-89
	unsatis.		54		54		74
Unfixed tops of sand ridges after preliminary fixation of the sand surface	good		80		60		80
	average	36	45-79	30	45-59	65	65-79
	unsatis.		44		44		64

Y_1: Accessible moisture supply in a 1-meter soil layer in early spring [mm]

10.6 Criteria for Prediction of Pasture Yield

Before sowing in different regions of the Karakum, it is necessary to assess the accumulated storage of accessible moisture in soil or DSW for various relief elements, and proceeding from the obtained results to take a decision on the advisability of sowing on them.

During an autumn-spring period, certain conditions of soil wetting are created, which make possible a correct appraisal of the sowing time of pasture plants.

Because of microclimatic differences, each element of the relief in an autumn-winter period is characterized by different amounts of precipitation, uneven DSW and accessible moisture supply in soil, which are important factors for adaptability of seedlings.

A pre-sowing period is appraised by accessible moisture supply, DSW or the amount of precipitation; it is necessary to estimate a sum for a period from November through early spring. Table 10.2. gives average monthly values for accessible moisture supply, DSW, precipitation in relation to different adaptability of seedlings in various elements of relief.

Table 10.3. Agrometeorological indices for various relief elements in the Central Karakum and their assessment for adaptability of seedlings (in the sowing period from spring to May)

Relief element	Adaptability of seedlings	Precipitation March- May [mm]		Y_1		Limit of WSD*	Root system depth in Aug. [cm]
		Average	Varying	Average	Varying		
Contact zones of takyrs and sands	good		80		50	80	90
	average	67	60-79	46	30-49	60-79	60-89
	unsatis.		59		29	59	59
Inter-ridge takyr hollows	good		90		60	90	90
	average	67	70-89	40	40-59	70-89	70-89
	unsatis.		69		39	69	59
Inter-ridge small hillocky sands and lower fringes of eastern and western slopes of a sand ridge	good		80		40	100	100
	average	69	60-79	30	25-39	80-99	80-99
	unsatis.		59	24	24	79	79
Middle of western and western slopes of a sand ridge	good		150		60	120	120
	average	67	100-149	18	40-59	100-119	100-119
	unsatis.		99		39	99	99
Unfixed tops of sand ridges after preliminary fixation of the sand surface	good		100		50	110	110
	average	63	80-99	20	30-49	90-109	90-109
	unsatis.		79		29	89	89

Y_1: Accessible moisture storage in late May [mm]
* in late May [mm]

An accessible moisture supply within a one-meter soil layer or DSW were estimated directly before sowing in a region of work performance. In the regions where a meteorological station did not carry observations over soil wetting or DSW these indices can be estimated as they are related to the amount of precipitation in November through February. Correlation coefficients of these relations are r = 0.76-0.88; 0.03-0.05, while a regression equation will take the form,

$$Y_1 = 0.85x - 5.1$$
$$Y_2 = 0.68x + 12.1,$$

where, Y_1 is an accessible moisture supply in a one-meter soil layer in early spring [mm] and Y_2 is DSW [cm]
x is the amount of precipitation in November through February.

About half of the average, annual precipitation occurs in spring. The significance of spring precipitation for adaptability of seedlings increases if the amount of precipitation in the autumn-winter period does not have an average value. Conditions of soil wetting in spring and their assessment under respective variations can be estimated using Table 10.2. The amount of precipitation, the accessible moisture storage in soil or DSW in pre-sowing and spring periods provide for different adaptability of seedlings. Lowering of precipitation in a pre-sowing or spring period makes the significance of a moisture content in the other period more critical.

In late spring the lower limit of a wetted soil layer becomes an important factor. Qualitative assessment of a lower limit of a wetted, soil layer in late May is given in Table 10.3. The influence of high temperatures of air and soil on seedlings depends on the depth of their root system. The greater is the depth of soil drying out in summer, the less the number of adapted plants. There is a danger of perishing for seedlings of plants the roots of which did not reach a 100 cm depth on slopes of a sand ridge and a 80 cm depth on small hillocky sands. A qualitative appraisal of adaptability of seedlings by a depth of their root systems in August is also given in Table 10.3.

In summer, seedlings perish within a one-meter, soil layer if moisture storage drops to 6 mm, and in the actual root layer if moisture storage is less than 5 mm. In summer in the contact zone of takyr-like hollows and sands and in takyr-like depressions, moisture storage in soil is not lower than 6 mm; on small hillocky sands and in lower parts (approximately one-third of total height) of eastern and western slopes of a sand ridge, soil moisture storage can drop below 6 mm on July 25, and in the middle part of eastern and western slopes of a sand ridge even earlier than in other elements of relief by June 30.

The proposed method which was elaborated in the Desert Institute can be used for agrometeorological assessment with differentiated sowing of ameliorative seeds in different elements of a relief in soils wetted to varying degrees. Therefore, the main criteria for prediction of yields and phytoreclamation works on pastures e.g. in the Karakum desert can be formulated as follows:

- Prediction of yields of natural pastures for the following year proceeding from data on cloudiness in September and October of the current year as an elementary index of the circulation in the atmosphere at the beginning of autumn, and the establishment of cyclonic types of weather in desert territories.
- In order to enhance the efficiency of phytoreclamation, it is necessary to forecast the expected level of seedlings adaptability.
- On the basis of the above forecasts, adoption of a strategy of differentiated decision-making concerning estimation of load on natural pastures, and the appraisal of advisability of phytoreclamation works in different years.

References

Arbykov K (1968) About prognosis of forage plants yield in depending of water infiltration into soils. Probl. Osv. Pustyn, 1968, No. 6, p. 80-82.

Fedoseev AP (1964) Climate and pasture grasses of Kazakstan. Gydrometeoizidat 1964.

Gringof IG (1965) To the prognosis of low grass yield on sandy pastures in Kyzyl-Kum. Rational utilization of desert pastures. Tashkent: Nauka, p. 89-94.

Nurberdiev MN (1976) About longterm agrometeorological prognosis of grass yield in Karakum. Problems of Gidrometeorology of Turkmenia. Ashgabat: Ylm, p.42-47.

11. Modeling the Development of Desert-Pastural Cattle Breeding

V.N. Nikolaev, V.V. Nikolaev, M.D. Nepesov

ABSTRACT: The problem of maximizing fodder reserves, one of vital importance to Turkmen farmers, is addressed. Using computerized analysis with the help of new software has significantly advanced our work in this field. However, even the earliest publications of the Desert Institute were very helpful in understanding the problem. The parameters necessary to sift the complex, data mass are listed. numerous variants for sheep breeding in different seasons and also in combination with camels are assessed. The conclusion suggests that the best combination is sheep and camels and that the sheep should not be on the pastures all the year round.

Low fodder productivity resulting from the climatic conditions, as well as the excessive cattle grazing, determine the great variability of the forage reserve status; under such circumstances, the periodic accounting and controls are necessary. Together with this, the detailed inventory making of natural fodder resources, in addition to the great organizational difficulties and the considerable material expenditures involved has led to difficulties in this domain for some years. Therefore, importance was always attached to the problems of systematization of the data of the complex, geobotanic examinations of pastures.

However, the processing of the huge quantity of data was carried out without the application of computers. In extreme situations this deficit led to an inability to intervene in the processes of agricultural production, as well as influencing the processes, occurring in the natural surroundings as an effect of the anthropogenic influence which have resulted in the degradation of natural meadowlands.

11.1 Breakthroughs Using Computers

More recently increased possibilities of using computers have permitted new and modern approaches to the problems of systematization, processing and analysis of the state of natural meadowlands of Turkmenistan and their utilization.

In this article, the newest achievements in the field of management systems of data bases of the relevant sort have permitted uniting in the framework of this unique functional medium, data, obtained over many years by different research and systematized in various standards. This variety of standards in many cases had resulted in their isolation from each other, and the subsequent impossibility of performing integrated processing.

All the available materials, concerning the complex examination of desert pastures of Turkmenistan have been analyzed by us. Despite both the early date of the publications and the existence of other, more recent cartographic materials and documents, we take the greatest interest in the monograph "*Natural Fodder Resources of Turkmenistan*" (Nikolaev VN 1972) from the Desert Institute of Turkmenistan, the "*Map of Pastures of Turkmenistan*" on a scale 1:800 000, and Nikolaev VN 1976) and the "*Recommendations for Improvement of Utilization of Pastures of Turkmenistan*" (Nechaeva NT *et. al.* 1977).

The choice of a data base does not imply dogmatic adherence to that or other information. The data systematization, their storage and replenishment, the possibility of making selective characteristics for any components, which are necessary to decide concrete problems in production activities of pastoral farming, give the possibility of conceiving the situation visually, and to intervene in the degradation processes on the pastures as well.

Modern information technologies offer a broad range of programs and hardware for EDP. The choice of one or another method depends first of all on the character of the data itself and the principle of their processing.

The presentation of information in the form of a text, large amounts of statistics, and graphic images would be a simple and typical approach. The newest hardware with multimedia capabilities enable the processing of audio and video information as well.

In our researches the information is offered in the form of statistics with textual explanations, by which the state of the object under investigation is described. The results of processing of our data may be offered by the same textual and statistical methods, which describe the object state also in conditions modeled by investigators.

The modern state of industry of the EDP successfully eliminates some problems of a labor consuming character during the information processing about the objects of investigations. On the one hand, the established idea of "machine modeling" has resulted in the appearance of a whole spectrum of computer programs, oriented to the creation of machine models of objects of investigations and the possibilities of multivariate decisions. On the other hand, the universal processing programs give us the possibility of selecting the most suitable form of data processing, reducing under it the labor-consuming character of some processing stages. And finally, the development of more or less sophisticated programs gives still greater possibilities in the presentation of information not only in the form of numbers and texts but in the form of complex, graphic images.

11.2 Choice of Software Program

The choice of one or another means and method of processing of data is based on a number of criteria. In this article the following criteria have been taken into account.

The final choice has been made out of three multifunctional, instrumental program methods: *Open Access* III/IV; *Microsoft Office* v. 4.3; *Novell Lotus Notes* V. 3.0. The first two packages are universal, integrated packages with approximately the same functions: the possibility of processing textual and tabular information, as well as the data, organized according to various principles. These two packages differ in qualitative development: the first functions in the DOS framework, the second in WINDOWS with all the resulting additional possibilities and advantages. The third package is a powerful means of processing and sending of data, organized in any formats and distributed by the computer network.

We took into account cost and accessibility as the main criteria. We prefer either the "Integrated System of Processing of the Information Open Access IV" of the firm SPI (USA), working in the DOS framework and permitting the utilization of even the simplest personal computers of the type AT/PC 286, or the package. "Microsoft Office v. 4.3" from the firm Microsoft (USA) to work under a Windows framework with the utilization of more modern personal computers of the type AT/PC 486, 586 or other *Pentium* chips.

11.3 Analyzing the Turkmen Data

In the process of information analysis of the state of natural meadowlands, the structure of the data base "Natural Fodder Resources", presenting the list of attributes/parameters with them abbreviated or with full names and correlations, has been created.

The information available nowadays about the state of natural meadowlands as a result has been systematized and introduced into the data base; this information is a totality of data, characterizing minimal, marked out contours on the map of pastures of Turkmenistan (scale 1:800 000). To these data have been added such parameters, as the territorial boundaries for the existing administrative divisions, natural geographic pastoral areas, the larger pastoral units and their anthropogenic modifications with the data of irrigation (in percentages) per farmed area, per cent, general and edible fodder reserves along with their nutritiousness, both for seasons of the year and annual averages.

All the starting point information of the objects of research is a mass of numerical data with the textual support so that one can interpret it, localization and some signs. The input of information into the system may be carried out both by means of textual editor and tabular processing. There is always the possibility of data conversion from one processing means into the other one.

11.4
The Development of Models

The model we developed has complex data permitting a timely intervention in the degradation processes occurring in pastures, to predict their state and to model, in accordance with the set tasks, various variants of the desert-pastoral, cattle breeding development.

The control implementation for the state of desert pastures depends on the indices and characteristics which are introduced into the data base, because the means of information processing permit the making of a choice in any criteria rapidly. So, for example, in the data base there is a field "Mod-type" which refers to anthropogenic modification of the type of pastures. This means, that these large, land areas have already undergone degradation. By means of machine processing, they can be easily separated with all the initial and calculated indices.

Proceeding from the indirect signs, the choice may be made. Irrigated pastures, for example are constantly undergoing repasturing while the unirrigated pastures on desert massives are overgrown with moss forming the crust, which doesn't create space underneath it to be developed for grassy plants. The sand areas desert are more liable to wind erosion, but mountain slopes and near-by plain surfaces suffer rather from water erosion.

Specialists, who work in the field of environmental protection, know many direct and indirect signs, which one or another way are connected with data of the informational base. The methods and ways of the restoration of degraded lands are known as well.

As already mentioned above, the created data base may be added to, corrected, reorganized and interpreted by various methods. At subsequent data inputs, new methods of interpretation may be always introduced into the base and determine the correlations with already existing parameters.

11.5
Improvements in Breeding

Pastoral breeding is of great importance for Turkmenistan. Despite low crop capacity, the desert pastures due to the huge territories annually renew hundreds of millions of quintal of fodder.

Traditionally desert pastures are utilized the year round. Climatic factors promote this to a large extent. The problems of animals pasture arise in winter months in connection with general reduction of fodder reserve and meteorological conditions although the latter factor varies sharply from area to area.

The animal population should not be kept on pastures the whole year but two or three seasons only. The pastures capacity for generating fodder using this cycle will be sharply increased. In the rest of the year, the animals will have to be kept inside in the stalls; for this it is necessary to have an irrigated fodder production, which takes up a considerable part of the ploughed lands of Turkmenistan.

In connection with this, we undertook the automation of the calculations, directed to the increase of the cattle-breeding production under the condition of full irrigation and pastures mastering.

11.5.1 Calculations of Breeding Possibilities

So, the calculation results of possible breeding development both for velayats and for Turkmenistan in a whole have been given in following variants:

- sheep breeding production all year with sheep keeping on pastures (insurance reserve on pastures) (Table 11.1.);
- sheep breeding production all year with sheep keeping on pastures (without insurance reserve on pastures) (Table 11.2.);
- sheep breeding production during spring, summer and autumn with sheep keeping on pastures (insurance reserve on pastures) (Table 11.3.);
- sheep breeding production during spring, summer and autumn with sheep keeping on pastures (without insurance reserve on pastures) (Table 11.4.);
- sheep breeding production during spring and summer sheep keeping on pastures (insurance reserve on pastures) (Table 11.5.);
- sheep breeding production during spring and summer sheep keeping on pastures (without insurance reserve on pastures) (Table 11.6.);
- Camel breeding production throughout the year with camel keeping on pastures (without sheep and with sheep) (Table 11.7.).

11.5.2 The Best Combination: Sheep and Camels

This is a far from full list of possible variants of the breeding development of Turkmenistan, but the proposed variants are objective enough and sharply differ among themselves in their final results. Our conclusion is that the combined utilization of pastures for the pasture of sheep and camels is the most optimal.

The calculations of the animal population and breeding production were carried out, proceeding from the following indices.

During a year it is necessary to provide 9.5 quintal of an airy-dry fodder for one sheep taking into account the insurance reserve in the case of a fruitless year. If the insurance reserve is stored on the ploughed lands, one sheep will require 8.1 quintal of fodder a year.

Table 11.1. Calculation of possible sheep-breeding production all year with sheep keeping on pastures (insurance reserve on pastures)

Velayats	Pasture land [1000 ha]	Number of sheep [1000]	Sheep-breeding production			Ploughed land [ha]
			Meet [t]	Wool [t]	Karakul*	
Balkan	10,291.95	1,330.03	11,704.24	4,123.08	532,011	15,447.43
Akhal	9,065.41	1,202.14	10,578.79	3,726.62	480,854	10,976.53
Lebap	7,940.34	976.53	8,593.48	3,027.25	390,613	12,791.66
Mary	7,670.79	1,503.32	13,229.20	4,660.29	601,327	10,779.92
Tashauz	5,670.68	615.57	5,417.05	1,908.28	246,229	7,098.61
TOTAL	40,639.17	5,627.59	49,522.76	17,445.52	2,251,034	57,094.15

* pieces of fur

Table 11.2. Calculation of possible sheep-breeding production all year with sheep keeping on pastures (without insurance reserve on pastures)

Velayats	Pasture land [1000 ha]	Number of sheep [1000]	Sheep-breeding production			Ploughed land [ha]
			Meet [t]	Wool [t]	Karakul*	
Balkan	10,291.95	1,559.91	13,727.20	4,835.72	623,963	27,484.72
Akhal	9,065.41	1,409.91	12,407.23	4,370.73	563,965	19,357.15
Lebap	7,940.34	1,145.31	10,078.77	3,550.48	458,126	23,739.09
Mary	7,670.79	1,763.15	15,515.73	5,465.77	705,260	20,178.61
Tashauz	5,670.68	721.98	6,353.33	2,238.10	288,788	12,036.07
TOTAL	40,639.17	6,600.26	58,082.26	20,460.80	2,640,102	102,795.64

* pieces of fur

Table 11.3. Calculation of possible sheep-breeding production during spring, summer and autumn with sheep keeping on pastures (insurance reserve on pastures)

Velayats	Pasture land [1000 ha]	Number of sheep [1000]	Sheep-breeding production			Ploughed land [ha]
			Meet [t]	Wool [t]	Karakul*	
Balkan	1,029.95	1,614.83	14,210.50	5,005.97	645,932	18,755.23
Akhal	9,065.41	1,493.48	13,142.67	4,629.80	597,394	13,636.79
Lebap	7,940.34	1,173.05	10,322.85	3,636.46	469,221	15,365.88
Mary	7,670.79	1,897.36	16,696.78	5,881.82	758,945	13,605.51
Tashauz	5,670.68	679,74	5,981.70	2,107.19	271,896	7,838.55
TOTAL	40,639.17	6,858.46	60,354.50	21,261.24	2,743,388	69,201.96

* pieces of fur

Table 11.4. Calculation of possible sheep-breeding production during spring, summer and autumn with sheep keeping on pastures (without insurance reserve on pastures)

Velayats	Pasture land [1000 ha]	Number of sheep [1000]	Sheep-breeding production Meet [t]	Wool [t]	Karakul*	Ploughed land [ha]
Balkan	10,291.95	1,892.70	16,655.75	5,867.37	757,080	33,348.30
Akhal	9,065.41	1,750.47	15,404.17	5,426.47	700,189	24,032.84
Lebap	7,940.34	1,374.90	12,099.14	4,262.20	549,961	28,497.78
Mary	7,670.79	2,223.85	19,569.85	6,893.92	889,539	25,451.11
Tashauz	5,670.68	796,70	7,011.00	2,469.78	318,682	13,281.99
TOTAL	40,639.17	8,038.62	70,739.91	24,919.74	3,215,451	124,612.02

* pieces of fur

Table 11.5. Calculation of possible sheep-breeding production during spring and summer with sheep keeping on pastures (insurance reserve on pastures)

Velayats	Pasture land [1000 ha]	Number of sheep [1000]	Sheep-breeding production Meet [t]	Wool [t]	Karakul*	Ploughed land [ha]
Balkan	10,291.95	2,166.35	19,063.87	6,715.68	866,540	25,160.78
Akhal	9,065.41	2,175.29	19,142.59	6,743.41	870,118	19,862.29
Lebap	7,940.34	1,666.33	14,663.73	5,165.63	666,533	21,827.42
Mary	7,670.79	2,875.24	25,302.08	8,913.23	1,150,095	20,617.61
Tashauz	5,670.68	856.64	7,538.42	2,655.58	342,655	9,878.50
TOTAL	40,639.17	9,739.85	85,710.69	30,193.53	3,895,941	97,346.60

* pieces of fur

Table 11.6. Calculation of possible sheep-breeding production during spring and summer with sheep keeping on pastures (without insurance reserve on pastures)

Velayats	Pasture land [1000 ha]	Number of sheep [1000]	Sheep-breeding production Meet [t]	Wool [t]	Karakul*	Ploughed land [ha]
Balkan	10,291.95	2,543.11	22,379.33	7,883.63	1,017,242	44,808.10
Akhal	9,065.41	2,553.61	22,471.13	7,916.18	1,021,442	35,059.30
Lebap	7,940.34	1,956.13	17,213.95	6,064.00	782,452	40,544.98
Mary	7,670.79	3,375.28	29,702.44	1,0463.36	1,350,111	38,628.81
Tashauz	5,670.68	1,005.62	8,849.45	3,117.42	402,248	16,764.85
TOTAL	40,639.17	11,433.75	100,616.30	35,444.59	4,573,495	175,806.04

* pieces of fur

Table 11.7. Calculation of possible camel-breeding production throughout the year with camel keeping on pastures (without sheep and with sheep) (Nikolaev VN *et. al.* 1984)

Velayats	Pasture land [1000 ha]	Camel population [1000]		Camel-breeding production			
		with sheep pasture	Without sheep pasture	with sheep pasture		without sheep pasture	
				Meet [t]	Wool [t]	Meet [t]	Wool [t]
Balkan	10,291.95	462.36	240.69	2,589.24	1,248.38	1,347.88	649.87
Akhal	9,065.41	341.06	140.71	1,909.96	920.87	787.97	379.91
Lebap	7,940.34	261.56	98.80	1,464.73	706.21	553.30	266.77
Mary	7,670.79	329.51	78.95	1,845.23	889.67	442.14	213.17
Tashauz	5,670.68	194.68	91.68	1,087.96	524.55	513.42	247.54
TOTAL	40,639.17	1,589.17	650.83	8,897.12	4,289.68	3,644.71	1,757.26

* pieces of fur

In addition to that, any of the additional feed for sheep in winter months is required to cover the protein deficit and during those times when the animals are in stalls. The area of arable land for the production of fodder was calculated out of the total requirement of subfodder means. Utilizing the simplest fodder crop rotation (Nikolaev VN 1972):

- lucerne in the first year and covered barley;
- lucerne in the second year;
- lucerne in the third year;
- winter grains and sorghum for hay 591.5 tons of fodder units may be obtained from 100 hectares of arable land.

The camels population has been determined for the pedigree of arvan, proceeding from an annual requirement of airy-dry fodder of 57 quintal.

The sheep-breeding production has been considered for the astrakhan pedigree of sheep under the specific weight of female sheep in the herd of 80% (Annamukhamedova O 1984) and for the camel-breeding by the average indices of camel-breeding farming.

Depending on the source consulted the size of the sheep population in Turkmenistan, which may be kept, varies from 5.6 million head to 11.4 million head. The arable land for the irrigated fodder production under them will correspondingly vary from 57.0 thousand hectares to 175.8 thousand hectares.

650,840 camels may be kept together with sheep round the year on pastures. All the indices are considered for an average concerning the crop capacity a year. It is evident that if there are only two or three seasons of pasture utilization for the sheep pasture, farm buildings for the stall keeping are necessary.

11.6 Conclusion

- Of late, broader possibilities for the utilization of computers have permitted new and modern approaches to the problems of systematization, processing and analysis of information about the state of natural meadowlands of Turkmenistan and their utilization.
- The available information today about the state of natural meadowlands has been systematized and introduced into the data base. The data base may be added, corrected, reorganized and interpreted in various ways.
- The programs facilitate timely interventions into the degradation processes occurring on pastures, to prognosticate their state and to model, in accordance with set tasks, various variants of desert-pastoral, cattle-breeding development.
- As an example 8 variants of the pastoral cattle-breeding development of Turkmenistan have been given. In the variants the number of the sheep population in Turkmenistan, which may be kept under the condition of full irrigation and mastering varies from 5.6 million to 11.4 million head. The arable land for irrigated fodder production under them is variously estimated at 570,000 hectares to 1,758,000 hectares. 650,840 camels may be kept together with sheep all year round on pastures, All the indices are considered for an average by the crop capacity a year. For two or three seasons, in order to improve the utilization of pastures for the sheep pasture, farm buildings for the keeping sheep in stalls are necessary.
- The elaborated program-informational complex is for the planning bodies, projective and research organizations, working on the problems of cattle-breeding, fodder production and the protection of nature. This complex may be interpreted for any integrated packages in the operating *DOS* and *Windows* systems.

References

Annamukhamedova O (1984) Livestock system of the Turkmen SSR. Ylym. Ashgabat.

Nechaeva NT and Nikolaev VN (1976) A map of pastures of Turkmenistan in a scale 1:800,000. Desert Institute. Ashgabat.

Nechaeva NT, Nikolaev VN, Smetankina VA (1977) Recommendation on improvement of pastures's utilization of Turkmenistan. Ylym. Ashgabat.

Nikolaev VN (1972) Natural fodder resources of Turkmenistan. Ylym. Ashgabat.

Nikolaev VN, Darymova GN, Orazdurdyev S (1984) Perspectives of the development of camel husbandry using combined desert pastures. Problems of desert development, No. 5.

12. Plant Growing on Near-Oasis Sand

A.O. Ovezliev, O.P. Kurbanov, A.I. Ibragimov

ABSTRACT: We focus on the near-oasis areas which are quite extensive in Central Asia. We assess the origins of such areas, their reclamation and subsequent development especially for agriculture and plant growth. For decades now the Desert institute has been experimenting with different methods and the results are summarized. Several land farms in Turkmenistan have been started on their basis. There are grounds for optimism, and the capital costs should be repaid within 3-5 years.

For development of low-value bad lands and increases of crop production in the arid zone, the sandy terrains near oases located inside and amidst oases are of great interest. The total area of sands nearby oases in Central Asia is more than 5 million ha. From ancient times the local population used these lands for cattle grazing and firewood cutting. Their natural productivity is very low, in some places they are severely degraded and transformed into barkhan sands that threaten settlements, irrigation structures, irrigated lands and other important agricultural objects.

Meanwhile, as the tests by the Desert Institute of Turkmenistan for 25 years have indicated: through applying appropriate technology these sandy terrains can be an important reserve for farming development. And this has been proved in practice.

Near-oasis sandy terrains in Turkmenistan stretch along rivers, irrigated lands, around and inside settlements. The greater part of their area directly borders on irrigated lands. Quite often irrigated lands cut in sandy terrains, and sands encroach upon oases (Photo 12.1.).

Near-oasis sands were formed largely as a result of the deflation of ancient, alluvial sediments. An area of sandy terrains neighboring only on irrigated lands in Turkmenistan is more than 1 million ha. They are mostly represented by hillocky, barkhan-hillocky, barkhan-ridge and barkhan sands lying over loose, sandy, alluvial sediments, takyr-like soils, ancient irrigated lands of loamy and loam, sandy texture.

Photo 12.1. Near-oasis barkhan-hillocky sands before development

12.1 Physical and Chemical Properties of Near-Oasis Sand

The granulometric composition of the near-oasis sands is mostly small-grained. The content of physical clay in them (d<0.001) is not more than 8%, of a silty fraction 6%. The minimum, moisture capacity and maximum hygroscopicity are very low. Within a one-meter layer they make, respectively, 5-6% and 0.6-0.9% of the weight. Water permeability of sand is very high. During six hours of observations, the saturation capacity was 1100-1200 mm or 11-12,000 m^3/ha of water at a filtration rate 3.1-3.2 mm/min.

These sands are marked by a very low content of nutritive elements. In the upper 0-30 cm layer the humus content is 0.06-0.15%, that of nitrogen 0.008-0.022%. The level of mobile forms of nutritive elements in these sands is very low. There are only traces of mobile forms of nitrates. The content of mobile forms of phosphorus varies from 2-8 mg/kg, exchanged forms of potassium 80-95 mg/kg, and the amount of easily soluble salts is insignificant.

Near-oasis sands contain an insignificant amount of silty fractions, feature high water permeability, low moisture capacity, and are prone to deflation. Only integrated development, and proper arrangement of the territory aimed at curbing of deflation processes and enhancement of their potential fertility can make these terrains into highly productive lands.

12.2 Development of Near-Oasis Sand

In many arid countries of the world great attention is drawn to development of methods of plant development of near-oasis sands. For this purpose in the Libyan desert, Saudi Arabia, and Algeria fresh groundwaters are widely used for sprinkling irrigation.

A progressive method of plant development of near-oasis sands is also drip irrigation which is widely applied in deserts of Algeria, Israel, Kuwait, Libya, Mexico, and Tunisia. According to A.G. Babayev's estimations about 500,000 ha of desert lands in the world are developed applying this method (Babaev 1995).

In Central Asian countries with their acute deficit of fresh water, the plant development of sands is more often based on the use of slightly saline groundwaters and shallow-lying subsurface waters. According to Morozov and Ivanov (1968) as a result of development of the Kyzylkum desert lands, the dry mass obtained was alfalfa 130 quin/ha, Sudan grass 92 quin/ha, corn 184 quin/ha, and sorgo 212 quin/ha. In the Muyunkum deserts in Kazakstan after application only of mineral fertilizers, the yield of green mass of Sudan grass reached 500 quin/ha.

In Turkmenistan, the study of plant development on eolian sands was first started in the 1930s at the Repetek sandy desert station. The possibility of growing various agricultural crops under natural conditions and additional water supply was studied in different regions of the Karakum desert (Semevskyi 1955). However, the results of experiments were not steady, not always reliable, and sometimes even contradictory due to a poor methodological basis, the small scales and sporadic nature of experiments.

Purposeful and systematic researches on plant development of near-oasis sands were started in connection with construction in the 1960s of the Karakum canal. At that time they studied the possibility of plant development of near-oasis sands by introducing ornamental, fruit-bearing, and melon crops under limited water application and using shallowly occurring, fresh groundwaters. As a result they identified promising plant varieties and developed methods of their cultivation (Babaev 1973).

In the 1970s the Desert Institute undertook wide-scale and in-depth researches in order to develop plant development of near-oasis sands. They studied ecological peculiarities of near-oasis sands and methods of their planting. It was found that in their genesis, water-physical and chemical properties the near-oasis sands are quite suitable for growing their forage, for fruit-bearing and other agricultural crops. Typical features of water physical properties of sands are the prevalence of small-grained sand in their profile, low values of the minimum, moisture capacity, high water permeability, and of agrochemical properties one should note the very low content of nutritive elements, low buffer and adsorption capacities due to a low content of colloidal particles in them. However, with correct agrotechnical and hydrotechnical practices many kinds of agricultural crops can be grown here, in particular, annual and perennial cereals and legumes, fruit trees and vines.

While growing crops on these sands, the best method for irrigation of forage crops is sprinkling and of fruit trees and vines i.e. drip irrigation (Desert Institute 1991).

12.3 Agriculture on Near-Oasis Sands

In the early 1970s, the Desert Institute investigated introduction of new forage crops from countries of Asia, Africa and Latin America and studied in detail their bioecological peculiarities in specific conditions of a sandy desert under sprinkler irrigation. More than 1,500 varieties of forage crops were tested: of these most suitable to extreme conditions of the Karakum and the most high yielding were the following varieties of forage crops: perennial forage sorghum, millet, Guinea late sorghum, Kafir White sorghum, Kafir brown sorghum, Feterit sorghum, cowpea, and dolihos. Under appropriate agricultural practices, they can give a yield of green mass of 500-1000 quin/ha (1 quintal=100 kg) (Ovezliev 1984).

Agricultural practices, in particular water application rates, nutrition regimes, a thickness of sand, optimal crop rotations of corn, alfalfa and some introduced forage crops were studied in detail. The optimal water application rate of forage plants on sands varies from 250-300 m^3/ha, while irrigation rate from 6,000-8,000 m^3/ha. The most effective application rate of fertilizers for corn growing on sands is N-200 kg/ha, P-120, K-60, Zn-5, Cu-1 kg/ha and organic fertilizers (manure) – 10 t/ha. The best pattern for location of tilled crops is 60 x 30 cm with two plants in a bunch. A six-field, crop rotation is most efficient (Desert Institute 1991, Ovezliev *et. al.* 1986).

In 1981-1985 the Desert Institute carried out tests of 34 varieties of grapes on the South-Karakum station, including 14 varieties of tables grapes, 17 varieties of seedless raisin grapes and 3 varieties of technical grapes applying a method of ampelographic collection. These varieties were taken from the Turkmen test station in Garrygal.

Tests have shown that for development of the near-oasis sands, most promising are the following grapes varieties: Ak seedless, Volga-Don, Gozel-Kara, Kara-Palwan, Seedless rose, Kara uzyum Ashkhabadsky, Kibraisky, Oktyabrsky, Early VIR, and Chol. These demonstrated not only high tolerance to extreme conditions of a sandy desert, but gave a high yield (100-150 quin/ha) of quality grapes.

In 1981-1985 first tests with 8 varieties of seed and stone fruit plants (apple tree Krasny prevoskhodny, Djanatan, Golden prevoskhodny, plume-alycha hybrid Partizanka, pear Santo-Maria, quince Aromatnaya, apricot Arzami and late Konservny) obtained from the Gekdepin state nursery. Among them quince Aromatnaya, varieties of apple trees (Plate 12.1.), apricot late Konservny showed best adaptability, conservation, growth and vegetation, adaptation to near-oasis sands and yield. After relating a fruit tree garden to a category of a fruit-bearing garden, their yield was 120-212 quin/ha (Ovezliev *et. al.* 1986).

At this station the Desert Institute also introduced 25 varieties of a pomegranate. Under sprinkler irrigation, application of appropriate rates of organic and mineral fertilizers the pomegranate varieties Ferghansky late, Paper shell, Surkh anor, Kayum anor, Guleisha Azerbijanian, Vishnevy soft-seed VIR-6 and others showed the most intensive growth, development and the highest yield. Beginning from the age of six years with conservation at least 90%, the yield was 90-145 quin/ha.

Twenty varieties of almond selected mostly at the Nikitsky Botanical Garden (Russia), Turkmen Test Station in Garrygal, and some foreign botanical gardens were tested under the same conditions for introduction and study of biological peculiarities. The Turkmen otlichny, Krymsky, Neck Plus Ultra and other varieties showed good results. A yield of dry nuts beginning from the age of nine was on the average 2.5 kg per tree.

12.4 Methods of Plant Growing on Near Oasis Sands

As can be seen from an analytical overview of plant development of deserts, the greatest experience in this field was accumulated by the Desert Institute. This allowed development of practical recommendations on growing of forage crops on the near-oasis sand terrains in Turkmenistan. Taking into account local, soil/climatic conditions, this experience can be applied to other sandy deserts in arid countries.

Growing of plants on sandy terrains involves a complex of economic, engineering and organizations actions aimed at enhancement of land fertility, radical improvement of their water-physical, agrochemical and biological properties with a view to obtaining large and stable yields of agricultural crops.

The profitability of sandy terrain development becomes higher when applying scientifically valid, irrigation regimes, optimal rates of organic and mineral fertilizers, methods for the chemical and biological protection of plants from enemies and diseases.

A system of reclamation measures on plant growing on near-oasis sands includes the following technological processes:

- leveling of sand terrains;
- construction of up-to-date irrigation systems (sprinkling, drip, etc.);
- construction of roads and infrastructure;
- preparation of an area for sowing agricultural plants;
- carrying out of agricultural practices.

Recommendations on plant growing on near-oasis sandy terrains should be implemented consistently, strictly observing a sequence of technological processes. In order to avoid deflation processes, it is not allowed to start land leveling works before water delivery, and decisions have been made on its distribution over the developed plot. In the course of construction works, it is necessary to ensure permanent wetting of a surface of leveled sandy terrains. Construction and maintenance works should be carried out in succession without any interruptions. To lower water consumption it is advisable to undertake land leveling during a wet period of a year (November through March).

During land leveling works on sandy terrains intended for sprinkler or drip irrigation, it is necessary to wet from time to time the leveled surface. So first of all rough leveling is made. At this stage hillocks are leveled out and pits are filled in, thus, creating good access for machinery. Mobile units with flexible and movable pipelines can be used for irrigation.

After leveling, water-supply irrigation should be carried out to prevent deflation development.

A process of land leveling and their subsequent development for growing agricultural plants should be consistent. On areas intended for sprinkler irrigation after the end of leveling works, one should apply a mix of organic and mineral fertilizers with a ratio of 10 t/ha of manure, 150 kg of phosphorus and 50 kg/ha of potassium active component. Then a presowing cultivation is carried out with an overturning of ground to a depth of 27-30 cm. A sprinkling machine is installed and a water supply irrigation is carried out at a rate of 250-300 m^3/ha depending on natural wetting of sands; then forage crops are sown.

The optimal time for sowing alfalfa is autumn (October). It is sown under barley. For row crops, such as sorghum, corn and others the best time is spring when an average daily temperature of soil at a depth of seed embedding reaches 10-12°C which is observed, depending on a climatic zone in Turkmenistan in the second half of April or in early May. Taking into consideration specific, thermal-physical properties of sandy soils, their susceptibility to deflation, the best depth for embedding alfalfa seeds is 2.5-3 cm, corn seeds 5-6 cm, seeds of sorghum and Sudan grass 4-5 cm. A ribbon method is used for sowing. Spacings among rows of row crops are 60-70 cm, of alfalfa 7.5-8 cm. The sowing rate of corn is 40 kg/ha, sorghum and Sudan grass 15 kg/ha, alfalfa 20 kg/ha. Under favorable climatic conditions seedlings appear 5 or 6 days after sowing. After 5 or 6 leaves appear on plants, the seedlings are thinned leaving 2 plants of corn every 30 cm and sorghum every 20 cm.

In plant growing on sandy terrains it is necessary, simultaneously with sowing of pioneer crops, to plant forest shelterbelts of such trees as white acacia, elm, apricot, and others. Depending on a field area the forest shelterbelts should be planted by the perimeter of an irrigated field and along an irrigation system. The main four-row, forest shelterbelts are planted perpendicularly to the direction of prevailing winds every 150-200 m from each other. Inter-row spacings are 3 m, and intervals between trees in a row are 2 m.

When selecting sprinkling machines for irrigation of sandy terrains, preference should be given to sprinkling machines and units that take water from a buried irrigation network. In sprinkler irrigation it is necessary to take into account air losses of irrigation water that may reach in a desert zone 12% of an application rate. The greatest water losses are observed in a daytime when temperature and humidity deficit of air rise sharply, thus, making irrigation in that period useless. Evaporation losses also increase with the decrease of sprinkling intensity.

Considering the low content of nutritive substances (especially humus and nitrogen) in sandy soils, it is necessary for enhancement of their fertility regularly to apply organic and mineral fertilizers that do not increase the level of toxic elements in soil and yield. In view of the fact that developed sands are structureless, feature high water permeability, low moisture capacity and an insignificant content of a soil absorbing complex, it is not advisable to use high rates of organic and mineral fertilizers in one application. In this case the nutritive elements are quickly washed down to deeper layers and became inaccessible for plants.

In follow-on years, pre-sowing cultivation of soil should be done by boardless plows. Periods and depth of cultivation depend on the conditions of that year, the preceding years and the degree of development. In the first years of development the depth of sandy soils cultivation for row crops should be 20-25 cm, in subsequent years 25-30 cm. Cultivation should be carried within out regular periods. If sand is very dry, a presowing water-supply irrigation at a rate of 250-300 m^3 should be done, otherwise deflation sources appear on a field surface and a leveled surface of a field is broken.

In spring, sands get warmed earlier than loams, so there is less risk of forming a crust after precipitation. As a result, as soon as the required temperature gets established at a depth of seed embedding, the sowing of forage crops should be started.

The most important requirements for the growing of agricultural crops on sands are vegetative irrigation within the established periods and a proper water application method.

The required amount of water for sprinkling is estimated proceeding from a soil moisture content, a moisture deficit, evaporation losses and a depth of root zone in different stages of plant vegetation. Irrigation of forage crops is started when moisture content in an active soil layer drops to 65% of the minimum moisture capacity. In spring this is usually observed in 8-10 days, in summer 4-5, and in autumn in 6-7 days depending on the kind of cultivated crop. Before rooting, the thickness of a moistened layer should be 0.5-1 m. And it is this layer which is most subject to variations of moisture storage during irrigation, because more than 90% of roots of plants are found here. In this case the water application rate in sprinkler irrigation varies from 250-300 m^3/ha, and the irrigation rate 6,000-8,000 m^3/ha, which is 1.5-2 times lower than under traditional irrigation methods.

After sowing, the inter-row spacings should be treated only if there are any weeds. Deep loosening during cultivation in such conditions is not necessary.

Experiments conducted by the Desert Institute have shown that under sprinkler irrigation a yield of green mass of maize VIR-156 is 400 quin/ha, Dneprovskaya (maize) 200-500 quin/ha (Plate 12.2.), Uzbekskaya (maize) 100-600 quin/ha, sorghum Sivashskoye hybrid-4 476 quin/ha, Oranzhevoye 160-473 quin/ha, Turkmenskoye-1 928 quin/ha, lucerne of Iolotanskaya alfalfa-1763 125-150 quin/ha.

According to our estimations the capital costs, depending on the volume of leveling works and yields of forage crops, can be repaid in 3-5 years. This can be attained by mechanized and automated water application, diminishment of main treatment, reduction of a number of inter-space treatments, more rational use of irrigation water and increasing the efficiency of fertilizer application to sands without deterioration of produce quality.

Experience has shown that sandy terrains that were considered badlands earlier, can be turned into more valuable agricultural lands. Plant growing will ensure fixation of drifting sands, improvement of water physical properties and enhancement of the fertility of sandy soils.

Using the results of these investigations the Desert Institute prepared a document, '*Recommendations on Turkmenistan oasis sands agricultural development*'. It had been discussed and approved by the vegetation section of the scientific-technical board of the Ministry of Agriculture of Turkmenistan on July 18, 1991; the recommendations were published and are used in the project organizations and the country's economies.

On the basis of these recommendations, the Institute "*Turkmengiprovodkhoz*" attached to the Turkmenistan Ministry of Land Reclamation and Water has set up several large, cattle-breeding, state farms' projects in the oasis sands i.e. the Garabekaul cattle-breeding complex and the licorice growing *sovkhoz* (state farm) in Lebap velayat, *sovkhoz* "Sakhra" in Garagum etrap of Mary velayat, a model-experimental subsidiary economy in Ovadandepe of Akhal velayat and others with a total area of more than 10,000 hectares.

References

Babaev AG (1995) Problems of desert development. Ashgabat.

Babaev AG (1973) Oasis sands of Turkmenistan and methods of their development. Ashgabat.

Desert Institute (1991) Recommendations on agricultural development of Turkmenistan. Ashgabat.

Morozov NA and Ivanov VF (1968) Irrigation on the basis of underground waters. Ashgabat.

Ovezliev AO (1984) Introduction of fodder crops in Karakum. Ashgabat.

Ovezliev AO, Agaev A, Kurbanov OR (1986) Experience of fruit crops growing on near-oasis sands in Turkmenistan. Scientific-technical progress in desert. Ashgabat.

Ovezliev AO, Ibragimov AS, Semergey KI (1990) Fertilizing of corn in Karakum. Ashgabat.

Semevskyi BN (1955) Experiences of Repetek station on plant-growing in Karakum desert (1934-1938). Works of RPPS, vol. II. Ashgabat.

13. The Fixation of Drifting Sands

S.K. Veisov, V.P. Cherednichenko, I.P. Svintsov

ABSTRACT: A worldwide problem in desert and semi-desert areas is that of drifting sands. There is a long tradition in Russia going back to the 19th C. of research-based measures to combat this threat. The Desert Institute has over many years analyzed and classified different types of sand drift. In the article the most interesting methods of sand fixing are described. The method chosen needs to be based on local determinants. Phytoreclamation appears to be the best; also interesting in this context are the use of loose materials e.g. clay and gravel. A new method is explained using a liquid mixture of polymers and latex. Materials forming a filmy surface are another alternative.

One of the major challenges of the global desertification problem is the control of drifting sands. Sands encroach upon irrigated lands, settlements, roads, canals, engineering structures, causing enormous damage to the economies and social spheres of many countries. At present intensive industrial and agricultural development of deserts goes hand in hand with active construction of railways and motor roads, construction of oil and gas pipelines, and as a result natural landscapes with their specific soil and vegetation cover are destroyed and turned into mobile sands. The World Map (1987) shows that about 600 million ha of lands in the world are subject to deflation and turned into moving sands: 319 million ha in Africa, 116 in Asia, 137 in Australia and 4.6 million ha in South America. Considering the significance of problems of control of moving sand, many world countries developed special programs envisaging restoration of a vegetation cover, soil conservation and fixation of mobile sands. In the USA as early as in 1935 a special service was established to control deflation of sandy soils.

13.1 Historical Overview

Great experience has been amassed globally on fixation and afforestation of drifting sands. Phytoreclamation methods to control drifting sands are more traditional and widely practiced in deserts of Central Asia, Near East and in some countries in the north of the African continent. Wide-scale and regular developments in controlling sand drifts were started in the late 19th century in connection with construction of the Transcaspian railway. By the beginning of its construction the Russian afforestation specialists had enough experience to cultivate tree-shrub varieties on mobile sands in dry areas in European Russia.

Using this experience, by the end of the 19th century they had developed original and effective methods to protect railways from sand drifts in Central Asian conditions. Very important in this context were the works of Obruchev (1890) who proposed three categories of actions to control sand drifts:

- preventive actions aimed at maintenance of the existing vegetation cover;
- reclamation measures;
- protection measures aimed at reduction of sand drifts such as the clearing of railway tracks, construction of brows along embankment edges, etc.

Obruchev's proposals were used, developed and found their practical application in the works of the forest specialist Paletsky (1928) who for nearly 40 years studied the problem of control of sand drifts on railways in Central Asia, and later of Dubyansky (1928, 1954, 1957) and Khodjayev (1947).

In subsequent years scientists and specialists of the Central Asian Research Forestry Institute (Tashkent), the Desert Institute of Turkmenistan in Ashgabat, the All-Union Research Afforestation Institute, the Kazakh Research Forestry Institute and others undertook development of measures to protect engineering structures from sand drifts and deflation.

Much experience on moving sand control was accumulated in Turkmenistan, in particular at the Desert Institute which was a coordinator in this problem development. Sandy deserts cover more than 300,000 km² in Turkmenistan, of which approximately 17,000 km² are with drifting sands. The scientists of the Desert Institute developed and successfully applied a wide scope of measures to fix a sand surface: mechanical means, protection films of bitumen, crude oil, polymer materials, plant hedges, etc. Efficient and regular efforts on sand fixation on vast areas in Turkmenistan led to considerable diminishment of formation sources of drifting sands.

13.2 Classification of Drifting Sands

For many years the laboratory concerning drifting sands of the Desert Institute studied the problems of deflation, drifting and accumulation of sands, revealed major laws governing their movement and the formation of a sand relief, traced direct relationships between sand drifts and relief formation, developed a classification of drifting sands and the physical foundations of sand deflation.

There are different classifications of drifting sands that are a major criterion used in development of measures to fix sands.

For practical use, we can specially mention classifications prepared by Ratkovsky (1949, 1964) for Central Asian conditions, Petrov (1950, 1952) for mobile sands of the former USSR, and Fedorovich (1960) for sands of Western Turkmenistan.

A classification developed by Babaev (1973) takes into account the following factors:

- the thickness of eolian sand drifts;
- the relief forms of sands and depth of their breaking;
- the types and scales of sand migration;
- the ground which underlies sands;
- the origins of eolian sands;
- the depth of occurrence and salinity of groundwaters;
- water regimes and their dynamics;
- the physical and chemical composition of eolian sands;
- vegetation and vegetative regeneration of sands;
- the processes of breaking and natural vegetative regeneration of sands.

While developing efficient methods to control sand drifts and deflation, it is very important to study the physical foundations of deflation-accumulation processes in sandy deserts. Sand transfer with wind should be considered as the main process of deflation that forms eolian relief.

Mechanisms of formation and development of eolian forms of relief in sandy deserts and their classification have been studied in many papers. In the works on drifting sands these are papers by Znamensky (1942, 1950) and Ivanov (1989). Principal notions of a pulse theory of particles tearing off from a sand surface and their movement in an air flow were developed. This theory takes into consideration aerodynamic and impact impulses acting on a sand particle. As a result they derived a formula for the estimation of lift and movement of a sand particle in a suspended state that allows calculation of the height and length of a flight of a spinning particle depending on the state of a sand surface i.e. its elastic properties and wind speed.

Further investigations revealed two natural phenomena: uneven and cyclic transfer of sand with a gusty wind, and the micro-impact packing of sand particles by a wind-sand flow. As a result deflation (transfer) zones and accumulation (deposition) zones of sand appear. The phenomenon of micro-impact packings of particles leads to an increase of an angle of internal friction and, consequently, to the appearance of an angle of an originating barkhan form. It was said that a profile of an originating eolian form takes shapes that feature highly aerodynamic qualities.

In some cases types of movement and speed of sand migration are determined by agricultural practices used in plant growing and afforestation on sands. Sands moving transitionally demand a more intricate complex of actions: planting and sowing of sand-fixing plants in combination with various types of mechanical and chemical protection measures.

Fig. 13.1. Fixation of mobile sands with vegetation

Sands with an oscillating type of movement demand a lower number of protection measures which should be implemented on certain elements of relief subject to migration. The degree of their danger depends on the type and speed of sand migration.

In their design features the engineering structures can be divided into two types: linear and spatial. The first type includes motor roads and railways, pipelines, power transmission lines and others; the second type comprises open sites where different technological facilities (gas accumulation units, compressor stations and others), production buildings, settlements, cattle-breeding farms etc. are located.

Depending on the nature of an object the methods applied for its protection will be different. Thus, protection of a pipeline or a basement of power transmission, line supports assumes the prevention of sand blowing out from a trench or the basement of a power transmission line support, whereas protection of a motor road or a production site relies on prevention of sand accumulation on a road or site. In the first case, all efforts are aimed at control of the blowing out process, in the second case, a process of transfer and accumulation.

13.3 Fixation Methods

13.3.1 Protection with Plant Materials (Grasses and Shrubs).

Nearly a century of experience of control of sand drifts and blowing out from different economic objects has indicated that the most reliable methods of sand fixation is phytoreclamation i.e. vegetative regeneration of sands (Fig. 13.1.).

The idea of sand fixation by this method is to stabilize a mobile surface for some time (e.g. two to three years) providing a possibility for plants to root in there. For this purpose they resort to mechanical protection including planting of shrubs and grasses in rows, by squares, in the form of hedges or a mat. After stabilization of a sand surface (attainment of an equilibrium profile), they sow seeds or plant seedlings and cuttings of plants under this protection vegetation within the set terms and in conformity with certain norms.

By method of arrangements they are subdivided into creeping, semi-hidden and hidden. Besides, mat and standing mechanical protection methods can be continuous, in rows or square. The distance between rows and the side of squares are usually not more than 2-4 m. In squarely arranged mechanical protection, they often apply reed (cane), stems and twiglets of shrubs and semishrubs, as well as *Artemisia spec.*, *Stipagrostis karelini* and others.

Protection through local vegetation is most popular and its application has a long history. Possessing some positive features among which the most valuable are their low cost and excellent adaptability of sand-fixing plants (Babaev 1973), but they demand high labor input.

Semi-hidden mechanical protection methods have a number of advantages. Among their merits are a complete preservation of local vegetation and complete termination of sand migration and transfer from a protected site. Here favorable conditions are established for self-overgrowing of a surface layer. Application rates of protection material depending on the type of protection methods are the following: for reed (standing row protection) 100-150 m^3/ha, for mat protection of *Stipagrostis karelini* and camel's thorn 50-100 m^3/ha. The most reliable protection from moving sands is squarely arranged vegetation. However, as yet it requires much material and manual labor (Plate 13.1.).

A combination of semi-hidden protection with seed sowing or planting of seedlings of local plants is considered the best. For prolonging the lifetime of mechanical protection, it is necessary to undertake its repair in the first year of planting growth. In follow-on years, it is necessary to repeat planting of plants on sites where they did not root.

The best results are obtained from a combination of semi-hidden mechanical protection means with clay addition. Here mechanical protection means are arranged in squares sizing 2 x 2 or 3 x 3 m depending on wind regime and the state of a sand surface, while in an inter-square space, dry clay is spread in a layer 5-10 cm thick followed by its wetting and the simultaneous planting of psammophytes.

When twiglets of shrubs are used as protection, then it becomes unnecessary to level a sand surface. The existing vegetation and relief remain intact as a result. That is why this type of protection is most efficient because it completely puts an end to any migration of sand from a protected site.

One of the advantages of protection using twiglets of shrubs and semishrubs is that it secures seeds of plants in their places, so the thus protected plots overgrow quickly. Moreover, branches of cherkez, kandym and other plants placed during autumn, winter and spring can root in too. This type of protection should be assumed most suitable, although its arrangement requires much manual labor. Twiglets from shrubs should be cut manually and a distance not closer than 100 m from an object on a leeward side.

Facing a continuous expansion of construction in a desert, it is necessary to apply such types of protection that can maintain all positive features (good adaptability of plants, deflation control efficiency and suitable costs), and admit wide mechanization of work. It is also recommended to use protection through reeds. Due to its aerodynamic properties and installation methods, it is similar to protection of local plants, but it is more durable.

Usually drifting sands are fixed by installation of different types of mechanical protection means in combination with sowing and planting of sand-fixing plants. Mechanical protection means are considered as an interim measure to stabilize a barkhan relief for 2-3 years, and to protect seedlings and plantings which are sand fixing from blowing out and covering. For these purposes they usually apply squarely arranged protection means that are installed on eolian forms from a base to a top. Widths of protected zones vary from 100-500 m on their windward side and half that size on the leeward side.

Stabilization of a barkhan relief on a windward side is attained by arrangement of a protection strip 100 and 150 m wide for engineering structures going parallel to chains. In these places most effective are protection strips 1-1.5 m widely spaced every 3 m or arranged in squares sizing 3 x 3 m. Here windward slopes of barkhan chains are fixed from a bottom to a top.

13.3.2 Loose Protection Materials

Of loose protection materials the most popular are clay and gravel. They are placed either in a continuous layer (the most widely used method) up to 3-5 cm thick, or they are arranged in strips and squares. Such protection means are made manually, but these operations can be mechanized. Vegetation grows well on them. Such mechanical protection means were tested in Western Turkmenistan in 1991-1993.

Takyr clay can be also used for fixation of a surface of drifting sands. Application of clay to a sandy soil enriches it with nutritive substances necessary for the normal growth of plants and does not pollute soil.

Slopes of embankments of motor roads, power transmission line supports and water conduits should be strengthened by placement of a clay protection layer. Its thickness, depending on a particular object, varies from 10-15 cm. This method is not costly especially in a combination with phytoreclamation. Clay material can be usually found nearby, and its delivery does not require much cost.

Fixation of mobile sands by strips has its specific features. While building clay strips (ridges), it is necessary to sprinkle them with water to create a non-scouring crust. In some cases seeds of sand-fixing plants or seedlings of cherkez or *Calligonum rubens, C. setosum* can be planted in inter-strip spacings. The sowing rate of seeds of *Haloxylon spec.* is 6-8 kg/ha, that of cherkez, or *Calligonum rubens, C. setosum* 3-4 kg/ha, the rate of seedlings' planting is 3,300 pcs/ha. It is better to conduct works with dry clay in an autumn-winter period before the main rainfall period, or to wet clay by sprinkling it with water at a rate of 1.5-2 l/m^2. After wetting a clay crust to create favorable conditions for plant growing by protecting the surface from blowing out of dry clay by the wind and, consequently, blowing out of sand.

Loamy sands were used for fixation of sand surfaces in Tyuesu (Eastern Mangyshlak 1990-92). They made ridges across prevailing direction of winds or placed loamy sands in a layer 10-20 cm thick. Ridges were sprinkled with water at a rate 3-4 liters per m^2 of a surface (Jumashov *et. al.* 1995).

Laboratory analysis has shown that a clay content in samples reaches 30%. After sprinkling them with water a solid crust 2-2.5 cm thick forms on a surface of loamy sands (after drying). Investigations in an aerodynamic tunnel have indicated that crusts are not blown away at a wind speed of 7 m/s at a height of 30 cm from the surface. Cattle can break the crust but its additional wetting with water restores the crust strength.

A complex method of fixation of mobile sands envisages application not only of loamy sands, but of crushed gravel, too. It is also used as ridges 0.5 m wide and 0.1 m high. Protective rows of reed are arranged every 5 m with filling of loamy sand or crushed gravel on both sides of a reed row. Planting of seedlings or sowing of seeds are made on a leeward side of a row, that is in a wind shadow. In this case young plants will be protected from dominating strong winds. These protection measures will allow, firstly, the application of cheap, easily accessible and ecologically safe materials i.e. loamy sand or crushed gravel; secondly, saving of more expensive material e.g. reed; thirdly, increase of moisture content in a sand under a crust by reducing evaporation and, thus, improving adaptability of desert plants.

13.3.3
Binding or Liquid Materials

Of great interest is the protection of moving sands with liquid fixing materials. This kind of protection can be easily mechanized. However, not everywhere can one find a base for widescale application of this method. On the other hand, nearly everywhere many devices for application of this fixing material are available.

The most widespread kind of fixation of a moving sand surface is a mixture of petroleum products (90% oil, 8% fuel oil and 2% bitumen). A heated mix is prepared: first they heat bitumen, then fuel oil is added into the melted bitumen and after good mixing oil is added; after that it is sprinkled by a pump. A rate of a mix use is on the average 3 liters per square meter of a surface. Protected sites do not usually require preliminary and careful leveling.

Application of a protective mix of petroleum materials completely prevents blowing of sand from a protected site, but does not trap sand carried with wind from the outside in a wind-sand flow. To achieve that we favor the development of sand-fixing plants. Apart from petroleum materials, other chemical materials often being waste products can be applied to protect sand surfaces from being blown away.

Oil as a ready for use natural product is used in the fixation of drifting sands. Laboratory investigations have shown that an optimal rate of oil use for sand fixation is 2.5-5.0 l/m². As a result a crust 2-3 cm thick is formed. Experiments also proved that high-viscosity oil is not harmful for seeds. On an experimental plot of 20 ha treated with oil and sown with *Calligonum rubens/C. setosum* and motley grass, within two years a thick vegetation cover appeared that completely stopped sand drifts on a railway. Experiments with application of oil materials for protection of power transmission lines from blowing out were carried out by the Desert Institute between 1964-1978, and gave positive results.

In laboratory and field experiments, they tested as a binding material the oil containing 20-30% of resin and 1.97% of asphaltes. Coatings were placed not on a smooth surface, but on ridges 0.5 m high and 1.75 m wide. A steady coating was created at an oil application rate of 4 l/m².

The Desert Institute developed a method of fixing drifting sands by clay suspensions treated with polymers of a PAA group and latex. A field was irrigated with these suspensions at a rate of up to 4 l/m². Field experiments have shown that polymers completely prevent cracking of a clay crust after drying out, and it does interfere with growth of young seedlings. The mechanical strength of a clay crust with a polymer is 7-8 times higher than without it. Under a crust, a permanently higher, moisture content under a favorable temperature regime is observed. This creates good conditions for normal growth of plants. Unfortunately this method is not applied in practice for sand stabilization.

Beginning from 1964 the Desert Institute carried out laboratory and field experiments with polymers of the 'K' series, such as K-4, K-6, K-9. Experiments have shown that after treatment of sands with these polymers a solid crust is formed on the surface. Mechanical strength, wind erosion stability, rate and degree of polymer penetration into sand, and germination capacity of seeds were appraised. Under field conditions, one percent and lower concentrations of a working solution were prepared, and then applied to a sand surface using the amount of 1.5-3 liters per sq. m. Field experiments with application of polymers K-4, K-6 and K-9 on the Garagumy canal and in Western Turkmenistan have indicated the following: the protection layer so obtained is up to 20 mm thick, features high crushing strength (up to 10-12 kg/cm²), does not break when affected even by strong winds, does not interfere with seed germination and seepage of precipitation, and alleviates temperature variations under a crust in summer and winter.

Beginning from 1973 the Desert Institute carried out a series of laboratory and field experiments to create wind-resistant crusts by applying a solution of bitumen in waste motor oil and transformer oil. These oils that have been subject to repeated regeneration became unsuitable for further use (Plate 13.2.).

Studies of sand-protection properties of bitumen solution in waste motor oil and transformer oil (ration 0.5:10 and 1:10, respectively), and crusts obtained after application of these solutions permit the conclusion that these two oils can be successfully applied for protection of power transmission lines from blowing out. As the oil itself does not contain resins, it cannot create a crust with good winter-erosion stability, thus, it is necessary to add in it substances that contain resins and asphalt substances, for instance bitumen and fuel oil. In summer temperatures, variations under this crust are much higher than under a crust formed by polymers and clay, but much lower than under crusts of crude oil. These coatings are not harmful for plants.

In laboratory conditions, they studied viscosity variation of oil and technical lignosulfonates in relation to a temperature. They also determined the water-mechanical strength of crusts obtained by applying these materials. An optimal rate of use and concentration of oil and lignosulfonates for fixing drifting sands was found.

Laboratory studies of lignosulfonate have shown that the best properties were received with the crust formed as a result of sand treatment with a paste consisting of lignosulfonate and wood tar at a ratio 1:2 and its dilution with a ratio of 1:2, 1:3.

Laboratory studies of distilled purified oil (DOM) have indicated that its application rate should be at least 3 l/sq. m. However, it is toxic, thus, its application should be limited to sites where there are no people and cattle, and personnel working with this material should take all necessary measures identified as obligatory in handling harmful substances.

Experiments with a gas pipeline have shown that the optimal thickness of a protection crust under natural conditions is more than 5 mm. In this case a surface protection crust is steady enough and its lifetime can reach 2 or 3 years. That is why it is necessary to increase fixing material use to 3 liters per square meter, and not 800 g/m^2 as was practiced before. Such an application rate significantly enhances stability of coatings and their lifetime. As a result there is no need for frequent restoration of a protection complex. Correct performance of all works and increase of usage of a fixing material do not make it more costly, but on the contrary makes a sand protection complex cheaper, as it excludes pure earthen, repair works and expenses associated with multiple irrigation.

Adaptability of sand-fixing plants on sites where the surface is fixed with viscous materials depends to a great extent on their variety composition, planting terms, quantity of planting material, and the degree of the surface fixing. The best adaptability is observed with cuttings and seeding of *Calligonum rubens/C. setosum* (60-80%), above average adaptability with seeding of cherkez (50-55%), and the lowest with *Holoxylon aphyllum* (30-35%). Tests have proved that on sites where removal and accumulation are minimum, 90% of plants are maintained, while on unfixed sites nearly all plants perish.

Sand fixation applying liquid fixation materials allow mechanization of this process, which increases rates and scales of work, and as yet this is the main advantage.

13.3.4 Film Materials

A film material may be used as a mechanical protection method. Its application assumes installation of protection films reinforced by seedings of sand fixing plants by a mechanized method using serially produced forest planting machines. A film material (polyethylene, polyvynilchloride, etc.) 25-30 cm high is installed in sand with rows every 2-3 m with simultaneous planting of seedings of sand-fixing psammophytes one by one on each side. The lifetime of films is up to one year. Films are used for mechanical protection for a number of reasons. Films installed in soils are almost unaffected by chemical action of soil solutions and microflora. Their high mechanical strength and elasticity allow mechanization of its placement in soil in the assigned places and to a set depth.

Films as a mechanical protection means with simultaneous planting of cuttings or seedings of sand-fixing plants are installed across the direction of dominating winds arranged in strips every 3 m. Smothly rolling and levelled sands are subject to continuous treatment. In barkhan sands, protection is provided on windward sides up to 2/3 of their height. A reclaimed area in this case is not more than 60-70% per ha.

Using for protection purposes a polyethylene film 0.1 mm thick, its use per ha of physical area is 100 kg. The rate of planting material use (seedings or cuttings) is 3300 pcs per ha.

Seedings planted in contact with a film act as reinforcing material. Irrespective of the direction of dominating winds, they secure the upper edge of a film in a vertical state, thereby, forming an obstacle ameliorating deflation processes. Planting of seeding on barkhan sands by this method gives adaptability of plants from 31 to 87%.

Total costs of protection using film materials, according to Svintsov (1984) is 10 times less than arrangement of square, mechanical protection methods with the size 2 x 2 of reed mats, and 4-6 times cheaper than continuous fixation of the sand surface with viscous materials, such as oil, SSB, polymers, or narosine.

In 1971 a specific, non-traditional method was applied to protect the production site of the Nebitdag iodine producing plant. Taking into consideration very unfavorable conditions in the region (high salinity of underlying ground, strong and often recurring winds causing intensive transfer of sand and migration of barkhan forms), it was proposed to use a specific technology resulting from the factory: the availability of great quantities of wastewater. For this purpose several terraced lakes were designed on a windward side in front of the plant site. After they are filled with wastewater they became a form of trap for moving sand. Possessing considerable capacity they ensured protection of this plant for many years. When they are filled with sand, the embankments will be raised, and so the lakes' capacity will be increased to the required extent.

Such method for control of sand drifts on spatial objects is more suitable, because in these cases it is not necessary to provide costly protection means requiring high labor input. And it meets the principal demand of desert environment conservation: to restore a disturbed landscape around a site.

Efficiency of protection measures, irrespective of their location, directly depends on the chosen methodological approach. The methodology of protection measures should be based on careful study of natural climatic factors, organization of monitoring of deflation processes, the carrying out of experiments both in laboratory and field conditions for the selection of optimal and effective methods of fixation of drifting sands.

References

Babaev AG (1973) Oasis sands of Turkmenistan and methods of their development. Ashgabat. Ylym.

Federovich BA (1960) Problems of sand classification for mapping of samples of western Turkmenistan. Physical geography 5: p. 9-24.

Ivanov AP (1989) Formation of profiles of eolian relief forms of sandy deserts. Ashgabat. Ylym.

Jumashov AP *et. al.* (1995) Protection of engineering constructions on Mangyshlak Peninsula from sandy drifts and deflation. On the sample of sandy Tyuence massive. Probliemy osvoieniia Pustyn. 1995, 2: p. 83-86.

14. Plant Development on Takyr Soils

N.K. Lalymenko

ABSTRACT: Takyr soils which are widespread in central Asia, can be reclaimed for vegetation or even fruit growing. The Desert Institute researches, and especially those of the Nebitdag Afforestation station, are summarized. The takyr areas best suited for phytoreclamation need to be carefully chosen. Details of the recommended furrow system are given. Comprehensive data is required to estimate the exact size of the furrow a factor of key importance to their success. The furrows accumulate moisture and the precise advantages of this phenomenon are listed.

Takyr lands in the arid zones in the world are concentrated largely in drainless, accumulation landscapes on alluvial, proluvial, deluvial and anciently irrigated deposits (Plate 7.4.).

They are usually characterized by a flat relief, a heavy granulometric composition of top soil, a high salt content, very low water permeability, presence of water grass or lichens, and nearly complete lack of higher plants.

During rainfalls and sometimes during snow melting, a runoff is formed on takyr surfaces that from times immemorial have been used by the local population for cattle drinking, and the cultivation in small depressions ("oitak") of melon and other agricultural crops.

Physiological dryness of soils of takyr plains is the main reason why natural vegetation is lacked or very thinned. Applying century-wise experience of the local population and results of runoff investigations the Nebitdag Afforestation Station of the Desert Institute of the Turkmenistan developed a technology for plant growth on takyr soils.

All technological decisions of the Nebitdag Station are aimed at maximum attraction and rational use of local flow for water supply of cultivated varieties of cultural and native flora.

Our researchers have found that for plant growth on takyr soils (growing of tree-shrub vegetation and creation of pastures with motley grasses, growing of broad-leaved trees, nut trees and other orchards, vineyards and melon crops) using local surface runoff, it is necessary to apply a system of large furrows (trenches) that are cut perpendicularly to the main slope of a surface. Such furrows are the most primitive hydraulic structures intended for accumulation of local runoff and creation of moisture reserves in soil that will support production of cultivated plants. In this case, of primary importance are the shape of a cross-section of furrows, their configuration, a spacing between them (a watershed area of each furrow), runoff in the area of work performance, morphology of a vertical profile of takyr soils and others.

14.1 The Development of Takyrs with Help of Furrows

Of the numerous options that were developed and tested over many years, the trenches with a triangle cross-section (an internal angle 95-110°) with double-breasted summits that are made by trenchers proved to be the most successful (Plate 14.1.).

In the absence of such tools, furrows can be cut by single-furrow, large plows (trenching, marsh-shrub and others) in two runs. Furrows of such design display good stability to lateral scouring as the surface runoff gets into them through special water inlets that are built (with regard to a slope) in spoilbank crests every 50-100 m. An advantage of double-breasted furrows is that their filling with local runoff becomes a controlled process.

At an identical depth of single- and double-breasted furrows, the capacity of the former is twice as low as the latter. That is why for accommodation of the runoff matching the capacity of double-breasted furrows, it will be required to have twice as long single-breasted furrows, which will make an evaporation surface 60% more. Double-breasted furrows will allow the concentration of the local runoff and lower unproductive, physical evaporation due to diminishment of the total, wetted area of their surface.

14.1.1 The Estimation of Parameters of Furrows

In plant development of takyr lands, the following data are necessary to estimate parameters of furrows:

- The amount of surface runoff, calculated with the annual averages of many years, for values of daily precipitation provided by the nearest meteorological stations, and a runoff coefficient set for takyr soils. For instance, for regions in Turkmenistan where takyr soils are widespread, the average annual runoff varies from 122-170 m^3/ha.
- The average single runoff i.e. a frequently recurring runoff during one rainfall, or a sum of rainfalls following each other (estimations took into account only runoff of 50 m^3/ha and more). In Turkmenistan the average value of a single runoff of a high recurrence varies from 120-125 m^3/ha.

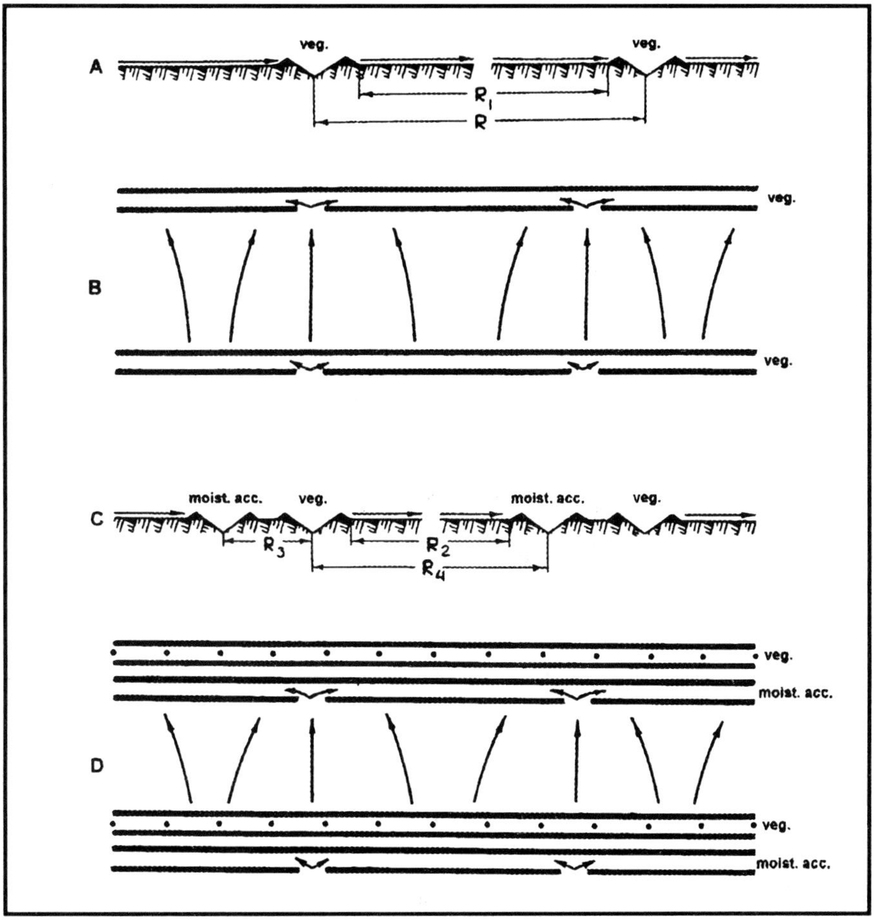

Fig. 14.1. Fragment Scheme of two marginal furrows on the surface of takyr soils: **(A)** vegetal furrows; **(B)** the same in plan view; **(C)** vegetal and moisture accumulated furrows; **(D)** the same in plan; **(R_1, R_2, R_3, R_4)** distance between furrows. The Arrows show the slope of surface and direction of flood. The Black dots are plants in furrows. The Splits in the thick black lines are water launching.

- Lithology of takyr soils in the area of work performance (granulometric composition of soils, specific alternation of layers of different composition, a depth of occurrence of underlying sands, loamy sands, etc.). Very different combination of these features are possible.
- A thickness of a layer that in its natural state is devoid of physiologically accessible moisture.

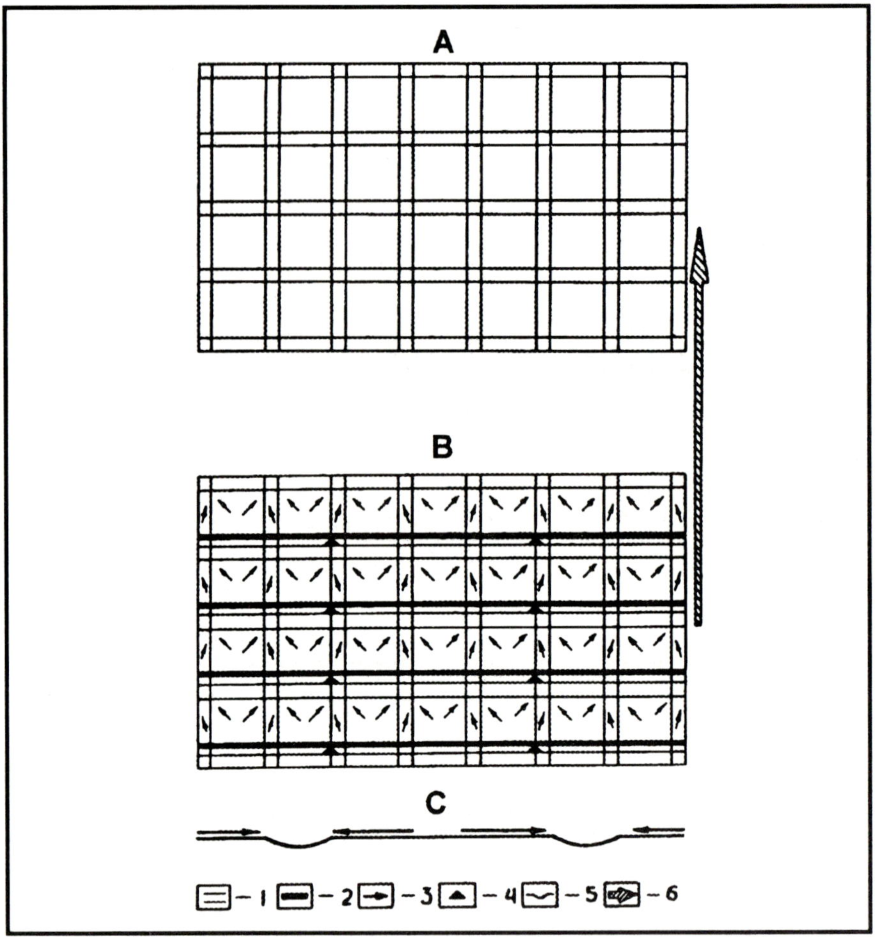

Fig. 14.2. Scheme of artificial, hydrographic, net device in interfurrow spaces and main furrows **(1)** scheme of a special road-roll's movement **(2)** furrows for sowing and planting **(3)** direction of surface flowing **(4)** water launchings **(5)** run of edge road-roller or car's wheel **(6)** slope of takyr ground

- An area of wetting contours under furrows ensuring the required for plant growth moisture content in soil.
- The amount of local surface runoff necessary for getting wetting contours to a depth of 2-2.5 m. On heavy (for their granulometric composition) takyr soils, a depth of moisture penetration can be 1.5-2 m; such plots can be used for cultivation of the most, drought-resistant plants.
- A time period necessary for formation of the assigned wetting contour under furrows. On takyr soils in Turkmenistan this process lasts for 2 or 3 years.

The above-mentioned criteria can be helpful for theoretical validation of estimation of initial parameters for plant development of takyr soils in relation to different geographical points in the arid zones.

14.1.2 The Use of Runoff

When growing plants on takyr soils relying upon the local surface runoff, it is very important to select an optimal layout of furrows. Investigations have shown that runoff can be used most rationally if each furrow has a separate watershed area (inter-furrow spacing where runoff is formed) (Fig. 14.1.).

With the diminishment of a runoff formation area (in our case inter-furrow spacing), losses of runoff to filling microdepressions and saturation by takyr soils decrease as the time for water movement from a watershed to a place of its accumulation (furrows) becomes shorter.

A considerable increase of the runoff coefficient (by 25% on the average) is provided by advance (before cutting of main furrows by a trencher) creation on the takyr surface of an artificial hydrographic network using a special heavy roller with welded ribs every 2-2.5 m ribs which have a triangular cross-section, or using a loaded truck (Fig. 14.2.).

In all cases this job should be done over a wet soil surface which provides for a smooth transition from a surface of takyr soil to a constructed groove.

At first on the takyr surface they cut single, double-breasted furrows ("vegetation"). Some time later – this time period depends on a complex of factors –one more furrow of the same size is cut near each of them; this process is called "moisture accumulating".

Runoff from a surface of inter-furrow spacing gets into furrows and wets soil in the form of "suspended" wetting contours, at first under vegetation, and in subsequent years under moisture accumulating furrows (Figs. 14.3 and 14.4.).

On the average in 2 or 3 years (sometimes within one year depending on a precipitation regime and a composition of a soil profile), the volume of wetted soil under furrows can reach several thousands of cubic meters per hectare (Figs. 14.3 and 14.4.: A, B). At the same time water soluble salts are leached from soil (under furrows). Over 7-10 months, the content of chlorides to a depth of 2-3 m can by reduced by 28-90%, and sulfates by 20-66%.

Within the periods set for a particular region, cultivated plants are placed in vegetation furrows. While growing perennial plants (fruit, nut and forest trees, grapes, etc.) in the 3rd or 4th year, it is necessary to cut a moisture-accumulating furrow near each vegetation furrow. Water inlets should be quickly made in crests of spoilbanks of second furrows.

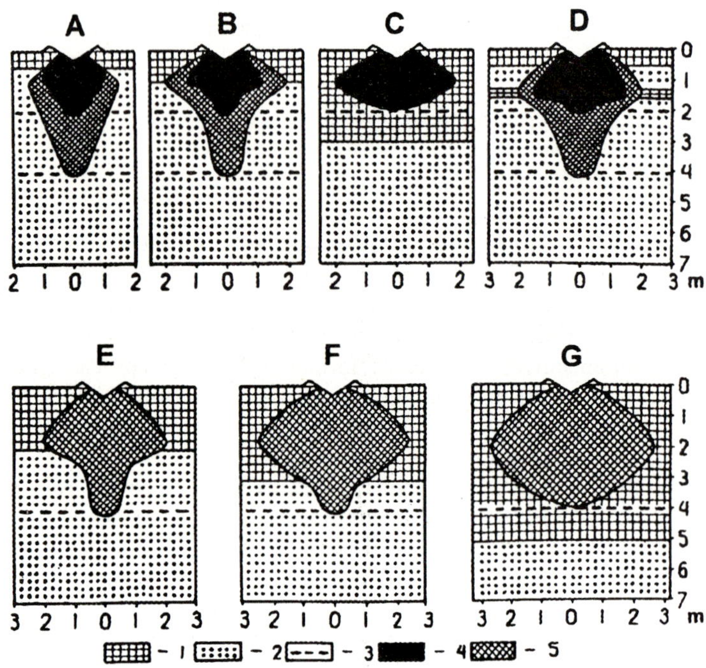

Fig. 14.3. Outlines of soilground soaking under two-part profile furrows of physically dry soilground early till 2 m or 4 m depth on takyr soils of different composition. **(1)** soilground of heavy composition **(2)** sand **(3)** lower boundary of impermacid layer (2 and 4 m) **(4)** outlines of soaking under capacity of impermacid layer of 2 m **(5)** the same under 4 m

14.1.3 Why Furrows Help

For growing *Haloxylon aphyllum*, seeds should be sown into moisture accumulating furrows immediately because seedlings of this variety die if flooded by water.

Moisture-accumulating furrows help resolving the following tasks:

- Closing up of soil wetting contours under vegetation and moisture-accumulating furrows (Figs. 14.4.: C, D, E, F) which can, in fact, double the volume of wetted soil.
- Under the influence of surface runoff, the volume of soil leached of water soluble salts increases. This creates favorable conditions for plant cultivation, opens

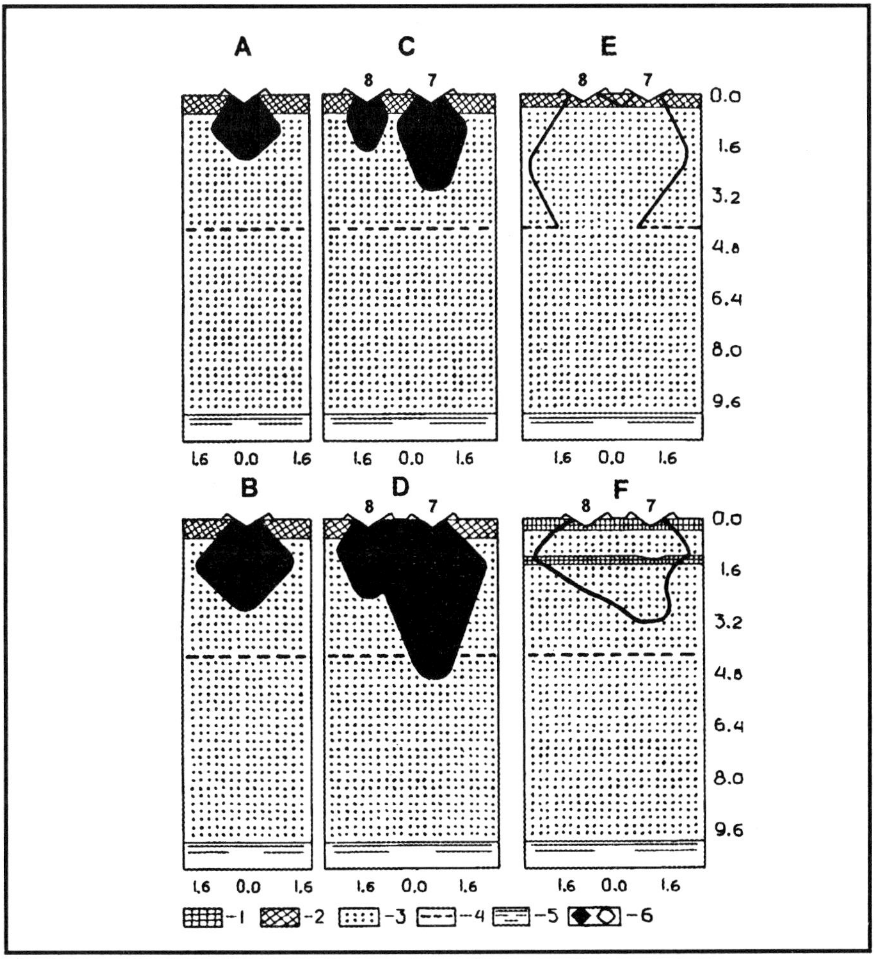

Fig. 14.4. Soaking of takyr soil under isolated, single and doubled furrows of two-part profile. **(A)** under two-parted furrow of 30-35 cm depth with R=13.5 m **(B)** the same with R=23.5 m **(C)** under two parted coupled furrows, with depth of 30-35 cm, under R=13.5 m **(D)** the same with R=23.5 m **(E)** the same with R=13.5 m for period **(F)** the same under R=13.5 m **(1)** clay **(2)** loam **(3)** loose sand **(4)** lower boundary of impermacid layer **(5)** subsoil water **(6)** outlines of soaking **(7)** vegetative furrows **(8)** moisture accumulated furrows

possibilities to grow new varieties of plants, and to put into economic use the takyr-like soils with a higher salt content.

- Water from surface runoff percolates to root systems of plants with all resultant positive appraisal of this water application method.

- A real possibility is created for regular removal applying technical means (a trencher) from moisture-accumulating furrows of a colmatage layer formed by sedimentation of silty particles being in suspension in surface runoff, this allows restore original filtration parameters of soils.
- In the course of development, root systems of cultivated plants tend to get into wetting contours under moisture-accumulating furrows where surface runoff restores from time to time a soil moisture content.

For melon crops (melons, water melons) and other annual plants, the moisture-accumulating furrows are not cut at all, because in this case there is a possibility that after harvesting they can be rehabilitated by trenchers, and as a result of the removal of silt deposits the water permeability of soils can be restored.

14.1.4
The Length of Furrows

In Turkmenistan, an average, surface runoff from takyr soils measured over many years varies within wide ranges between 122-170 m^3/ha a year. This makes it necessary to make differentiated estimations of all parameters of furrow cutting taking into consideration the above natural and experimental data. For instance, while creating shrub-grass pastures using desert flora, the length of furrows can vary as follows:

- heavy-textured soils down to 0.5 m underlain by sand or loamy sand (700-1000 lin m/ha);
- heavy-textured soil down to 1 m underlain by sand or loamy sand (500-1000 lin m/ha);
- heavy-textured soil down to 2 m and more underlain by sand or loamy sand (300-700 lin m/ha);
- interbedding heavy-textured horizons, loamy sands and sands (500-1000 lin m/ha).

Depending on furrow length on a unit area, all other parameters of furrow cutting are estimated (Table 14.1.).

14.2
The Tree-Shrub Varieties

In Turkmenistan for creation of pastures on takyr soils, it is recommended to sow or plant in furrows seedings (in polyethylene containers) of tree-shrub varieties *Haloxylon aphyllum, Salsola paletzkiana, S. richteri, Tamarix ramosissima* to which on the second year of vegetation the following varieties are undersown: *Salsola sclerantha, S. subaphylla, S. gemmascens, S. arbuscula, Suaeda arcuta,*

Table 14.1. Recommended parameters of furrows on takyr soil for bushes and grasses planting

Length	Depth	Width on the ground surface	Cross-section	Distance (s. Fig. 14.1.)				
				R	R_1	R_2	R_3	R_4
1000 m/ha	0.30 m	0.84 m	0.125 m^2	10	8	5.6	2.3	6.5
900 m/ha	0.31 m	0.88 m	0.138 m^2	11	9	6.6	2.4	7.5
800 m/ha	0.33 m	0.94 m	0.156 m^2	12	10	7.6	2.4	8.6
700 m/ha	0.35 m	1.00 m	0.178 m^2	14	12	9.5	2.5	10.6
600 m/ha	0.38 m	1.09 m	0.208 m^2	16	14	11.4	2.6	12.4
500 m/ha	0.41 m	1.19 m	0.250 m^2	20	18	15.3	2.7	16.4
400 m/ha	0.46 m	1.33 m	0.312 m^2	25	22	19.2	2.8	20.4
300 m/ha	0.54 m	1.54 m	0.416 m^2	33	30	27.0	3.0	28.3

Gamanthus gamocarpus, Halimocnemis karelinii, Petrosimonia drachiata and others. Besides, furrows are gradually silted with many other varieties of xero-halophyte flora making valuable forage whose seeds drop in furrows during winds.

For creation of pastures, suitable takyr soils are those with any composition of a vertical profile which in their natural state feature a salinity of a root zone (0-6 m) not more than 3.0-3.5% of a dry residue and 0.8-1.0% of chlorine. At a higher salt content, *Haloxylon aphyllum* is replaced by *Tamarix* or still more salt-resistant varieties typical of this region.

Growing of *Salsola paletzkiana* is possible only on takyr soils with a thickness of the top, loamy-clay layer not more than one meter, and underlain by sands, loamy sands or interbedding horizons of a heavy, medium and light granulometric composition.

When plants (*Haloxylon aphyllum, Salsola paletzkiana and S. richteri)* intended for cattle grazing, reach 4 years of age they should be cut or broken. After this operation, they give many new seedlings with a great number of assimilation twigs and small seedlings making a valuable feed for cattle. Subsequent cuttings – it is better to make them by a special cramp mounted on a tractor – should be made every two-three vegetation seasons in a cold period (December-February). Cutting of protection plantings is not recommended.

A system of vegetation and moisture-accumulating furrows opens wide possibilities for creating on takyr soils not only pasture and protection agrocenoses, but also plantations for growing more water-loving plants (fruit and nut trees, grapes, melon crops, broad-leaved forest trees, and others).

For successful management of this task it is necessary to select regions with takyr soils that have the most favorable precipitation regimes capable of providing an average, annual surface runoff of 150 m^3/ha and more.

14.3
Parameters for Water-Loving Plants

In this cases parameters of furrow cutting will be as follows: a length of double-breasted furrows 300-400 lin m/ha with R=25-30 m; R_1=22-30 m; R_2=19.2-27 m; R_3=2.8-3.0 m; R_4=20.4-28.3 m; (where R is the distance between furrows); a furrow depth 0.46-0.54 m; a capacity of one linear meter of furrows 0.312-0.416 m^3.

After wetting of soil under furrows to a depth of 1.0-1.5 m (Fig 14.3.: A-G and Fig. 14.4.: A, B), special sowing or planting holes are made in their floors in every 2.5-3 m with the following parameters: h=1.2-1.3 m; h_1=0.3-0.4 m; h_2=0.7-0.8 m.

Manure mixed with soil removed from this very hole during drilling are applied to the floor of holes (h_2) at a rate of 8-10 kg. One can add to this mixture 60 g of nitrogen-containing mineral fertilizers and 100 g of phosphorus ones. The upper part of holes (h_1) is filled with pure soil. If in the course of making this hole an underlaying sand horizon is penetrated, then loam or clay is added (20-30% by volume) to improve granulometric composition of a filling material in the upper part of holes (h_1). After filling, a mixture should be slightly packed.

After wetting contours under furrows going down to a depth of 2.5-3 m (Fig. 14.3.: A-G, Fig. 14.4.: A, B), sowing or planting of the selected varieties of plants should be made within the designated optimal period for the region. Before this, furrows should be cleaned by trenchers of all silt deposited from surface runoff which lowers water permeability of soils. At the same time, water inlets made in crests of spoilbanks of furrows should be restored.

14.4
Experiences with Cultural Plants

On takyr soils in Turkmenistan with episodic furrow irrigation by surface runoff, good results were obtained by growing of some varieties of cultural flora (Table 14.2.).

When growing *Haloxylon aphyllum* in twin furrows on takyr soils, by the end of the fourth year of vegetation it gives 15-18 quin/ha (1 quintal=100kg) of raw wood and up to 12 quin/ha of edible green mass. Besides, grass plants from local flora self-seeding in moisture-accumulating furrows give a yield of green mass from 10-15 quin/ha varying from year to year.

In the third year of fruition a yield of *Vitis vinifera* (wine) reaches, depending on varieties and water availability in a year, 6-8 quin/ha, while *Cydonia oblonga* (quince) reaches 12 quin/ha.

Pistacia vera (pistachio) start fruition from the ninth year of vegetation, and from the age of 15 it forms a yield of nuts and reaches 1-2 quin/ha.

Other fruit, nut and forest tree-shrub varieties intended for different economic use were subject to testing and they showed good conservation, satisfactory growth and fruit-bearing (Table 14.2.).

Table 14.2. Basic initial indicators of takyr soil for planting cultivation

Plants	Depth of upper loamy-clay layer	Permissible initial salt content (before forrowing)			Preparation of planting pits	
		Dry remainder	Cl	SO_4	with manure and mineral fertilizers	with mineral fertilizers only
Ulmus premila	1.5 m	1.8%	0.4%	0.8%	-	+
Morus alba	1.0 m	1.5%	0.3%	0.6%	+	-
Elaeagnus orientalis	0.8 m	2.3%	0.4%	1.0%	-	+
Maclura aurantica	0.8 m	2.3%	0.4%	1.0%	-	+
Gleditschia triacanthus	0.8 m	2.3%	0.4%	1.0%	+	-
Zizyphus jujba	1.0 m	2.2%	0.3%	0.5%	+	-
Amygdalus communis	1.5 m	2.0%	0.3%	0.8%	+	-
Pistacia vera	2.0 m	2.2%	0.3%	0.6%	+	-
Cydonia oblonga	1.5 m	2.5%	0.7%	0.8%	+	-
Punica granatum	2.0 m	2.2%	0.3%	0.6%	+	-
Vitis vinifera	2.0 m	2.5%	0.7%	0.8%	+	-
Citrullus vulgaris	1.5 m	2.5%	0.6%	0.9%	+	-
Melo sativa	1.5 m	2.5%	0.6%	0.9%	+	-

Yields of *Citrullus vulgaris* (water melon) varied from year to year and by varieties from 10-30 quin/ha, and of *Cucumis melo* (honey melon) from 12-15 quin/ha.

The above method of takyr land development by cutting furrows has various advantages: apart from additional provision of the population with plant growing produce and improvement of a forage base of cattle breeding, it has a positive ecological effect, because it allows the transformation of enormous amount of physical evaporation of runoff into transpiration of plants.

15. The Reclamation of Saline Soils

P.E. Esenov, K.R. Redjepbaev

ABSTRACT: Soil salinity is the worst hindrance to crop-growing; the researches of the Desert institute in this field are summarized. They are placed in the context of the work done or needing to be done in the five Central Asian republics. The specific, local, geographical factors are sketched in. The traditional method of leaching has both advantages and disadvantages. Of crucial importance for leaching is good draining. A more modern method is that of crop-growing with salt-friendly crops such as rice. The problem of soil salinization is complicated by difficult environmental problems such as water pollution.

In hot and dry climates, the main limiting factor for obtaining optimal yields of agricultural crops is soil salinity. So reclamation measures should be aimed at control of a salt regime of irrigated soils and enhancement of their fertility.

Natural factors on which the salt content of these soils depend are precipitation, evaporation, transpiration and groundwater salinity, original salinity of soils and some others. In conditions of the desert zone a permanent, upward flow of soil moisture dominates; this contributes to maximum, salt build-up on the soil surface.

In oases the above natural factors include the following: additional water applications, leaching, excessive water, seepage from irrigation canals, transpiration of cultivated agricultural crops, treatment and fertilizers (Kovda *et. al.*).

15.1 Salt Regimes

When assessing a salt regime of irrigated soils, consideration is given mostly to a coefficient of seasonal accumulation that reflects a ratio between a content of water soluble salts in a root zone in autumn and spring. There are three types of salt regime that are typical of non-irrigated and irrigated soils. In the first type of salt regime, a degree of seasonal soil salinization is approximately equal to seasonal desalinization, while in the second type seasonal salinization exceeds seasonal desalinization, and in the third type seasonal desalinization exceeds seasonal salinization. All three types of a soil salt regime are dependent on a complex of natural and economic conditions and their combinations. Therefore, water soluble and physiologically harmful salts that accumulate in the root and arable horizons are responsible for inhibition of agricultural plants; special measures are required to control a soil's salt regime.

15.2 Methods of Controlling Salinity

Actions to control soil salinity include the following: the application of crop management practices; land reclamation methods of desalinization of irrigated lands by carrying out leaching and leveling in autumn and winter; the growing of crop-developers of saline soils; new progressive methods of irrigation and drainage; the improvement of the existing agricultural practices; the application of efficient crop rotations; and optimal norms of application of organic and mineral fertilizers etc. These actions should be corrected in relation to particular, soil-reclamation conditions.

Under radical improvement of saline lands, leaching and drainage are widely applied. Their necessity is dictated by low yields of agricultural crops. In this context reclamation measures are intended, first of all, for medium- and heavily-saline irrigated lands, and newly developed soils with a different degree of salinity.

15.3 Research on the Effects of Salinity

Experiments of the Desert Institute of Turkmenistan have shown that on slightly saline soils (the sum of salt content in a water extract is not more than 0.5-0.6%), cotton yields drop to 15%, on medium-saline soils with a salt content 0.5-0.7%, they fall to 40%, on heavily saline soils (with a salt content of 0.8-1.0%) to 80%, and on very heavily saline soils (salt content is higher than 1.0%) practically no yield is obtained. Irrigated lands in Turkmenistan are saline to different degrees; as a result every year cotton yields lose from 5 to 12 quin/ha (Redjepbaev 1991) (Plate 15.1.).

In Central Asian countries saline irrigated lands cover large areas. As of 1989 the area of saline soils in percent to a total area of irrigated lands is: in Uzbekistan (51.4%), in Tadjikistan (15.4%), in Kyrgyzstan (11.5%), in Turkmenistan (89%). In general, the state of lands in the Aral Sea basin gives rise to serious concern (Table 15.1.). Detailed information about the specific features of desalinization measures in different regions of Central Asia are given in the next chapter.

Table 15.1. Distribution of irrigated lands in Central Asia by a salinity level within a 0-100 cm layer as of 1989 (Pankova and Solovjev 1995)

Salinity level	Turkmenistan		Uzbekistan		Tadjikistan		Kyrgyzstan	
Total registered area [1000 ha; %]	1219.1	100	4144.2	100	689.7	100	1034.2	100
including:								
non-saline	134.1	11.0	2012.1	48.6	583.4	84.6	915.2	88.5
slightly saline	458.6	37.6	1296.2	31.3	69.0	10.0	67.1	6.5
medium saline	471.9	38.7	598.7	14.4	28.7	4.2	36.0	3.5
heavily and very heavily saline	154.5	12.7	237.2	5.7	8.6	1.2	16.0	1.5
TOTAL of saline lands	1085.0	89.0	2132.1	51.4	106.3	15.4	119.1	11.5

15.4 Specific Regional Features

15.4.1 Turkmenistan

On the territory of the Akhalsky velajate in Turkmenistan it is necessary to take consideration the availability of large slopes of irrigated fields on the Kopetdag piedmont plain. Unlike other regions leachings here are made by feeding water via furrows 60-90 cm wide cut across a slope. A furrow length depends on a slope. After leaching, a field is leveled and prepared for sowing of barley, wheat or forage crops (Redjepbaev et. al. 1988).

In the modern Tedjen delta (Turkmenistan), lands in the old zone of irrigation became saline (Plate 15.2.), so farms started development of virgin lands beyond the cultivated zone. The existing drainage network (20 lin m/ha) is not capable of diverting heavily saline (22-24 g/l) drainage waters.

On irrigated lands in the zone of the 2nd stage of the Garagumsky canal (especially to the north of the canal), a complex reclamation situation was established that requires undertaking reclamation measures of all grades of difficulty, including leveling (400-2000 m^3/ha), leaching (3000-12000 m^3/ha) and drainage (from 20-30 to 75 lin m/ha).

In the Balkansky velajate in Turkmenistan, of the total area of irrigated lands equaling 62,190 ha more than 53,000 ha are in need of desalinization improvement. However, at present water of the Atrek river cannot be used for this purpose because of a high level of magnesium and a risk of magnesium alkalization of soils in the course of leaching. When waters from the Amudarya river come to the Atrek delta via the canal Gazandjik-Gyzyletrek, it will become possible to undertake desalinization.

Table 15.2. Area of saline irrigated lands in Turkmenistan requiring reclamation improvements as of 1995

Velajates	Irrigated area*		Salinity level* medium		heavy and very heavy	
Akhalsky	492.71	28.78	302.60	17.88	73.72	4.30
Balkansky	62.19	3.63	47.99	2.80	5.28	0.31
Dashkhovuzsky	401.92	23.48	281.59	16.45	66.45	3.88
Lebapsky	285.50	16.68	135.59	7.92	11.47	0.67
Maryisky	469.52	27.43	161.10	9.41	59.55	3.48
TOTAL:	1711.84	100.00	928.77	54.26	216.47	12.64

* in 1000 ha (first column) and % (second column)

In the Dashkhovuzsky velajate in Turkmenistan water resources in autumn and winter are limited. Irrigation waters starts flowing in early spring, and in this case leaching is often combined with water-supply irrigation. As a result soils at the beginning of vegetation remain unleached, and such fields are used for sowing of salt-resistant, development crops: sunflower, Sudan grass, sorghum, barley, millet and wheat. And only on the second or third year, the major crops can be cultivated here – cotton and corn. In the Dashkhovuzsky velajate, saline lands are reclaimed through rice growing. Here prolonged flooding of fields with a rate of 15,00-20,000 m^3/ha alternates with periodical discharges of water into a drainage network, or establishing a through-flow regime in the amount of 20-25% of the water supply volume. As a result already in the second or third year after water begins flowing, there are adequate conditions for cultivation of forage crops and even cotton.

The area of lands in the Dashkhovuzsky velajate that require leaching is 348,000 ha (Table 15.2.). The rate of leaching water application is differentiated, depending on a level of groundwater occurrence (UGV), salinity and texture of soil. On medium saline soils the area of which is 281,600 ha, leaching should be carried out at a rate 2,500-6,000 m^3/ha. On heavily and very heavily saline lands on an area of 66,450 ha, leaching irrigation is carried out at high rates 5,000-10,000 m^3/ha).

In the Lebapsky velajate in Turkmenistan, an area of 147,100 ha demands urgent land improvement. This is more than a half of an irrigated area in the velajate. The poor state of irrigated lands is connected with a high level of groundwater occurrence (34.3%) and soil salinity. Newly developed lands on the Yulangyzsky terrain in the south-west of the velajate are most difficult to improve. They require heavy leaching and improvement of drainage waters diversion.

In the Maryisky velajate in Turkmenistan the area of saline lands requiring urgent improvement is 220,650 ha. In the zone of old irrigation, about 40% of irrigated lands are affected by medium, heavy and very heavy salinization. For their improvement it is necessary to carry out heavy leaching, and to lower the level of groundwaters. In the velajate less than one-tenth of an irrigation network is provided with a filtration-proof lining, while the remaining part includes earthen canals, so seepage losses from them are very high.

On the Khauzkhansky terrain in the Karakum canal zone about 75% of irrigated lands are subject to medium, heavy and very heavy salinization (Redjepbaev and Esenov 1987).

Therefore, two-thirds of irrigated lands in Turkmenistan are in need of urgent improvement. Apart from leaching of soils it is necessary to construct new collection and drainage networks and repair those already existing, to reconstruct an irrigation network, to undertake radical leveling, and to improve water supply.

15.4.2 Uzbekistan

In Uzbekistan, irrigated lands are subject to active salinization. Areas of saline, irrigated lands in the republic make up on the average more than 50% (Table 15.2.). Taking into account that the irrigated land is not very useful, the priority task here is to improve the state of irrigated lands by way of their desalinization. The processes of soil salinization reveal themselves differently in different irrigated regions in Uzbekistan (Pankova and Solovjev 1995, Kamilov 1985).

In the Southern Pre-Ural (Karakalpakia), the prevailing processes of desertification and land degradation reveal themselves in salt build-up, intensification of dry, salt-dust transfers from a dried bed of the Aral Sea, extension of xerophyte and halophyte communities, and the shaping of new forms of the eolian relief.

In the irrigated zone of Karaklpakia, soil salinity is caused by heavy rates of water application, inadequate natural and artificial drainage, generally high salinity of irrigation water in the lower reaches of the Amudarya, presence of small-spot salinization and the brokenness of meso- and microreliefs. Therefore, improvement of irrigated lands here is connected with control of a water-salt regime of soils, enlargement of flooding checks, improvement of leaching and land leveling, and others.

In the Khorezm region, groundwaters occurring close to the surface and salt build-up in some localities have been observed. Under inappropriate, water application systems and the poor performance of drainage, a seasonal accumulation of salts takes place. In general, the Khorezm region is characterized by prevalence of a desalinization process that depends largely on good performance of a collection and drainage network, and quality leaching of soils every year.

In the Zarafshan river delta (Bukhara and Karakul oases), desertification processes are connected with the state of irrigated lands. In view of an impeded natural outflow, groundwaters here are characterized by a high salinity (5-10 g/l and more). Irrigated areas feature small-spot seasonal salinization of soils, and their provision with drainage is two times lower than the norm. So leaching in autumn and winter are obligatory here for improvement of lands, as well as the necessary wise management of water resources. For improvement of the state of irrigated lands, it is necessary to construct here an adequate drainage system taking into consideration specific structures of different parts of delta: vertical in the Kermine-Kanimekhsky and Bukhara oases and horizontal in the Karakulsky oasis. In terms of land reclamation measures, it is advisable to perform chemical improvements (gypsum treatment) to eliminate soil alkalization (a regional feature of soils in the Zarafshan valley), and apply better agricultural practices for improvement of hydro-physical properties of soils.

In the Karshi oasis located in the Kashkadarya valley, only the lower part of the delta is developed. Subsurface drainage is poor, processes of salt accumulation are connected with a hydromorphic regime of irrigation in combination with poor drainage. Groundwaters are heavily saline (more than 10 g/l). Land development in the Karshi steppe involves significant volumes of engineering and construction works (drainage) on areas difficult for improvement. Regular leaching of saline irrigated lands is also required.

In the Sharabad oasis the desertification processes have not yet established themselves, especially in the upper part of the Sherabaddarya river delta. In the middle and lower parts of the delta, salt accumulation processes are developing locally due to a low efficiency of a drainage network and an excessive water application rate. Considerable areas of takyr-oasis irrigated lands are subject to weak, in some places medium and partially heavy salinization. Cotton yields here are high (22.5-28 quin/ha) (1 quintal=100 kg) in comparison to other regions. In one of the major irrigated areas in Uzbekistan, in the Golodnaya steppe land improvement efforts that included application of a leaching regime in combination with optimal drainage resulted in desalinization of originally saline soils. Application of waterproofing of an irrigation network and higher irrigation rates allowed in the initial period of development (during 7-15 years) desalinization of soils and originally saline groundwaters (Kamilov 1985). On this terrain, and especially on newly developed lands, phenomena of spot and seasonal salinization of irrigated lands are widespread. Here it is necessary to control a water-salt regime by carrying out preventive leaching, lowering of irrigation rates to design values of water duty, observance of differentiated and optimal irrigation and drainage regimes.

15.4.3 Kazakstan

In Kazakstan slightly saline irrigated lands cover 311,400 ha, heavily and very heavily saline lands 1,112,500 ha (Kharin *et. al.* 1993). In the Kazakh part of the Aral region a share of saline soils in the brown soils is 55%, in the gray-brown and takyr-like soils 46%, and in the saz belt of piedmonts only 5%.

A delta of the Syrdarya river entering a province of chloride-sulfate salinization was subject to great alterations as a result of long irrigation. If earlier (before the Aral crisis) the soil and land reclamation situation was favorable, then in the recent decade as a result of a decreased water flow into the lower reaches of the Syrdarya an area of saline lands increased as well as salinity of river water. That is why a vast territory of the Eastern Aral region suffer from progressive desertification, consequently, it is necessary to undertake a complex of land improvement measures in irrigated regions.

15.4.4 Kyrgyzstan

In Kyrgyzstan soil salinization formed in various lithological and geomorphological conditions (Mamytov 1986). In floodplains and river terraces where groundwaters occur near to the surface, saline soils are characterized by a heavy surface salinization. In inter-mountain depressions and syrt uplands in places of outcrops of salt-bearing rocks, the lithogenus solonchaks are formed. With a relatively low amount of precipitation and high evaporation, automorphous saline soils of a deluvial origin are widely developed. In places where saline groundwaters comes out to the surface – which is observed mostly in intermountain depressions and valleys – hydromorphic, saline and solonetz soils are widespread (Bajenov 1973).

At present in Kyrgyzstan, irrigation is provided on best lands whose extent exceeds 1 million ha. In general, the position of land reclamation in Kyrgyzstan is positive due to good, natural drainage of its territories in comparison to other salt-affected regions in Central Asia. With time the area of irrigated lands will increase by including piedmont, low plateau and medium mountain areas, high intermountain depressions, and syrt uplands. In view of the specific nature of land salinization here, desalinization measures carried out in the course of irrigation should be aimed at: prevention of inflow of saltwater masses from the lower lying territories; drainage construction; gypsum application; application of higher rates of phosphorus and organic fertilizers; a leaching irrigation regime; sowing of salt tolerant, development crops (sorghum, sweetclover, alfalfa, barley and others); leaching of heavily saline soils with follow-on rice sowing on them; and the application of physiologically acid fertilizers (superphosphate, ammonium sulfate, urea) on solonetz soils.

15.4.5 Tadjikistan

In Tadjikistan in conditions of a primitive, long-fallow farming, areas under old irrigated lands suitable for cotton growing reduced to 20-30%, while low-productive solonchaks, heavily saline soils and marshes covered up to 55% of the area of old irrigated lands. Cotton yields were on the average 10-12 quin/ha of raw cotton (Kezim 1985). As a result of large-scale land improvement works (construction of drainage systems, leaching, rehabilitation of an irrigation and drainage network, etc.), the area of non-saline lands in the Vakhsh valley reached 71%, of slightly and medium saline lands, 25%, of solonchaks, and on heavily saline lands 4%. Marshes disappeared completely. Therefore, in recent years a favorable situation has established in the Vakhsh valley and in other irrigated areas of the republic. Nowadays it has become possible to obtain high yields of agricultural crops.

15.5 Methods of Improvement

For improvement of irrigated lands in Central Asia, numerous approaches and methods from the Desert Institute of Turkmenistan were developed and applied: physical (deep plowing, deep soil loosening, sand application and shifting of horizons in a soil profile); biological (cultivation of alfalfa, green manure crops and development crops, application of manure); chemical (application of gypsum, calcium chloride, lime and sulfuric acid); hydrotechnical (leaching and drainage, and electro-reclamation).

Leaching is a traditional method of improvement of saline lands. However, it requires large volumes of irrigation water which in areas with poor drainage conditions leads to a groundwater rise and renewed salinization. Therefore, leaching should be done if horizontal open or closed drainage is available. It should be attended by physical and biological land improvement methods. This will enhance efficiency of desalinization measures, and decrease consumption of leaching water.

Leaching capacity of soils (salt removal efficiency) and, consequently, water use for leaching are determined not only by a salt level in a leached soil series, but by their composition, distribution over a soil profile, soil texture and compacting, absorbing capacity and composition of absorbed bases, etc.

The main requirements of drainage are the lowering of groundwater level below a critical level (2.5-3.0 m), and their diversion together with the salts dissolved in them.

The best time for leaching saline soils in Central Asia is late autumn and winter. At this time of the year, saline soils feature the lowest moisture content, and a low level of groundwater occurrence. This is conducive to more intensive removal of salts from soil. On the contrary, summer leaching is least effective because much water is lost to evaporation, and there is a risk of repeated salinization. Plots where leaching are planned should be cleared of shrubs, and should be carefully leveled so that elevations varied within ± 5 cm and not more. Then plantation plowing or loosening should be done to a depth at least 50 cm in order to break up the earth, and improve water permeability properties. Deep loosening enhances salt removal already after the first applications of leaching water by 40-45% Redjpbaev *et. al.* 1988).

After loosening a final surface leveling is done. Careful leveling ensures constant desalinization of soil over the whole leached area, helps avoid water stagnation in depressions, and suction of salts to raised spots.

Then taking into account relief and diversity of salt-affected spots, a field surface is cut into fields to be leached, divided from each other by ridges 0.5 m high. Depending on terrain conditions, the area of each field could be 0.25 to 1.0 ha. After a field has been completely prepared, leaching can be started.

15.5.1
The Process of Leaching

The first dose of leaching water is applied in order to cover a moisture deficit which is the difference between moisture storage within a one-meter soil layer under maximum field moisture capacity and a natural moisture content. In each particular case a moisture deficit is estimated by a trial method before leaching is started. But practice has shown that by this time it is equal to 1,000-1,500 m^3/ha. The amount of water corresponding to the maximum field moisture capacity can dissolve all salts that are contained within a one-meter layer and which are to be removed even at the highest concentration.

In the following two or three days, in order to remove salt solutions leaching is carried out by applying separate doses of water (2,000-2,500 m^3/ha). Each subsequent dose is applied after complete percolation of the previous one. Following this pattern of leaching, one can feed to a leached field during one autumn-winter period about 4,000-4,500 m^3/ha of water. With higher water application rates, leaching can last for 2 or 3 seasons.

15.5.2
Sequence of Salts Leached Out

While leaching water percolates down into soil, it gradually replaces water soluble salts from a leached soil series. First removed are the most readily soluble chlorides of sodium and magnesium, and magnesium sulfate whose solubility remains nearly constant at varying temperatures. Sodium sulfates, carbonates and bicarbonates are not so readily soluble, especially at low temperatures.

In the course of leaching, positive results are indeed accompanied by some negative phenomena i.e. diminishment of nutritive substances reserves, deterioration of a water-air regime, increased alkalinity and development of soil alkalization processes. Lysimeter investigations have shown that losses of humus in the upper (0-20 cm) soil layer are 25 to 47% of its original level. Losses of nitrogen are also high but the content of phosphorus and potassium changes but slightly. Therefore, directly after leaching it is not recommended to sow such crops as cotton. First of all development crops (perennial forage grasses) should be sown on these lands with application of increased rates of organic and mineral fertilizers. If there is no manure (the required rate is 10-15 t/ha) for increasing soil fertility, it is recommended to sow green, manure crops.

15.5.3
Possible Additional Measures

If high water application rates are used (more than 7,000-8,000 m^3/ha), leaching lasts for 2-3 years, and after that it is necessary to undertake additional measures to prevent repeated salinization during inter-leaching periods. For this purpose after a leaching season, fields are sown with salt-resistant crops, such as joughara, sunflower, and sorghum. During their vegetation it is necessary to maintain a leaching regime of irrigation by increasing vegetation water application by 15-20%. Plants give shadow to a soil surface, so lowering evaporation, and in the course of treatment of inter-row spacings, a capillary link of subsoil moisture with the surface is broken.

15.5.4
Newer Methods

A new technology of improvement of saline soils is desalinization of solonchaks by growing rice on them in the first 2-3 years. During this period, salt content in solonchaks decreases to a required level, and later other crops can be grown on these lands. Desalinization of solonchaks by rice growing will be successful if proper drainage is provided.

During land reclamation for desalinization purposes, they sow salt tolerating development crops, first of all barley, Sudan grass, millet and others. After land reclamation, in order to prevent secondary salinization, it is recommended to apply more progressive irrigation methods – sowing on furrow crest, sprinkler, drip and subsoil irrigation. Here it is advisable to use more up-to-date and highly efficient, sprinkling machines "Fregat", "Kuban", "Volzhanka" and others. Leaching and vegetative irrigation should be carried out by furrows to gradually eliminate flood irrigation under which enormous quantities of water are lost to evaporation and seepage, especially on coarse-textured soils.

15.6
The Work of the Institute

Consequently, analysis of results of many scientific researches of the Desert Institute of Turkmenistan on saline soil improvement in the arid zone has indicated that on a territory with primarily saline soils and those prone to secondary salinization, the success of land development and soil desalinization depends to a great extent on natural and land reclamation conditions and economic activities. It was demonstrated that development of lands poorly prepared in irrigation and drainage terms (i.e. without drainage, leveling, or earthen irrigation canals) with application of significant water application and irrigation rates leads merely to a rise of saline groundwaters, the development of land hydromorphisation, and the intensification of secondary salinization of irrigated lands. These processes are most intensive in conditions of "nomadic", long-fallow farming, slight and moderate in regions with inadequate drainage of alluvial-deltaic lands. Radical improvement of saline lands assumes various actions: the application of horizontal and vertical drainage; lowering of a level of saline groundwaters; establishment of a negative salt balance on a territory; rehabilitation of existing irrigation and drainage systems and construction of newones; the application of differentiated crop management practices; and land improvement measures taking into consideration soil properties; etc. Permanent soil, irrigation and drainage monitoring of irrigated lands is very important for successful implementation of the above measures. No less significant are the burning contemporary problems of river water pollution and the deterioration of irrigation water quality.

An integrated approach to resolution of these problems will be conducive not only to a radical improvement of irrigated lands and obtaining steadily high yields of agricultural crops, but to improvement of the natural, environmental quality in arid regions, too.

References

Bajenov NK (1973) Saline soils of Kyrgyzia and ways of their melioration. Kyrgyzstan, Bishkek.

Kamilov OK (1985) Melioration of saline soils of Uzbekistan. Fan, Tashkent.

Kezim PA (1985) About development of salt-marsh lands of Vakhsh vabley of Tadjikistan. Melioration of soils of USSR arid territories. Ilim, Bishkek.

Kharin NG et. al. (1993) Explanatory note to map of anthropogenic degradation of soils in Aral Sea basin. Ashgabat.

Kovda VA et. al. (1967) Melioration of saline soils. Melioration saline and salted soils. Nauka, Moscow.

Mamytov AM (1986) Geochemistry of saline soils and meliorative zoning of Kirgizia. Ilim, Bishkek.

Pankova EI, Solovjev DS (1995) Remote sensing methods of dynamics of irrigated land salinity for reconstruction of geotechnical systems. Probliemy osvoieniia pustyn: 1995, No. 3.

Redjepbaev K (1991) Changing of natural meliorative conditions in Karakum canal zone, ways of rational utilization and protection of lands. Referat of dissertation and protection of lands. Seminar paper based on dissertation. Ashgabat.

Redjepbaev K, Irashkevich I, Tolstolytkin IG (1988) Recommendations on washing out saline soils of Karakum canal zone. Turkmenistan. Ashgabat.

Redjepbaev K, Esenov P (1987) Changing of soil-meliorative conditions of Khauzkhan massive in connection with irrigation development. Ylym, Ashgabat.

16. Precipitation Use

M.A. Nepesov, T.P. Vitkovskaya, B.T. Kirsta

ABSTRACT: The attempt to harness precipitation has a millenia-long tradition; deserts do have both precipitation and water but very precise constructions are needed. Of special interest has been the attempt to build artificial watersheds with newly developed methods, or by enhancing the usage of natural ones. Takyrs, widespread in Central Asia, have been at the center of both traditionally and experimentally utilization for water collection. The use of the newer technologies is assessed. Subsurface storage is of most interest here. The Desert Institute has become a unique base for both experimental design and testing.

Problems of the proper use of precipitation have a long history. And it has not lost its scientific and practical significance at present for arid regions where accumulation and storage of precipitation remain the only source of fresh water. At the same time at various stages of development of the human species, relations to the use of precipitation as a source of fresh water were not simple. This is reflected in the process of development and improvement of techniques for collection, storage and use of precipitation. In the course of many centuries the local population developed many original methods and unique hydraulic structures that were used for many millennia by desert people for storage of precipitation collected from natural watersheds. This supported their main occupation which was distant-range cattle breeding.

Precipitation was also widely used in agriculture: oitak farming (Turkmenistan, Kazakhstan, Israel, Iran), growing forage crops (India, Australia), plant growth on takyrs and improvement of natural pastures (Turkmenistan, Uzbekistan, Kazakhstan, Iran and others).

Central Asia is one of the world regions with a water deficit caused by various factors: hot air temperatures reach 50°C, while temperature of a soil surface can be as high as 70°C; relatively low precipitation (about 100-300 mm a year) in some areas, and a long dry period from May through October in which practically no rains fall are typical of this region. But at the same time this region possesses enormous water potential which with an average annual rate of precipitation in places of 150 mm is as large as 300-500 km^3. These potential water resources are 5-8 times more than an average annual flow of such rivers as the Amudarya, and undoubtedly it is of great importance for the organization of the water supply on the basis of precipitation.

The Karakum desert possesses the least potential water resources. But at the same time its desert pastures annually receive in the form of precipitation up to 50 km^3 of water (an average annual precipitation rate is 130 mm). This is nearly 1.5 times more than the water resources of all rivers in Turkmenistan, including such big rivers as Amudarya (with a watershed on its territory), Murghab, Tedjen and Atrek. It is interesting to note here that in conditions of even distribution of precipitation throughout a year their monthly rate will be 4 km^3, the daily rate 120 million m^3 and hourly rate 5 million m^3.

These estimations prove that there is water in a desert but it is necessary to learn to use it properly. That is why for many centuries people lived in deserts, mostly bred cattle there, grew melon crops and grapes on rain-fed lands, and practiced small-oasis farming.

Less than 5% of the annual precipitation would be enough to support cattle breeding – an agriculture which can quickly be developed – and development of small-oasis farming in a desert. In this case guaranteed water supply should be based on valid and accurate data on precipitation for many years covering different groups of low-water and high-water years and including extreme years by water availability (maximum of the maximum and minimum of the minimum). Here precipitation is considered to be water resources which can be used as a source of fresh water proceeding from studies of their regime.

16.1 Research into Precipitation

A regime of atmospheric precipitation was studied using observation data of the Hydrometeorological Service for the period 1942-1990 provided by 42 meteorological stations.

Analysis of the obtained results has shown that a precipitation regime is controlled by synoptic processes typical of a given territory. Under the existing general regularity of synoptic processes, there is no any relationship between the amount of precipitation and their type. In the last hundred years no changes in synoptic processes due to the growing impact of engineering activity on the natural environment were observed, and will not be observed in the nearest 15-20 years. Consequently, no alterations in a regime of atmospheric precipitation for the period in question were observed. An exception is an interval moisture cycle that was formed in regions of irrigated farming development and construction of major open water objects. Here, an increase of precipitation by 30-40% in relation to the background level is observed in some places.

Very important here are extreme values of annual precipitation that should be taken into consideration in development of a project, and the validation of capacity parameters of structures in order to create guaranteed, water reserves. This is connected with a specific regime of atmospheric precipitation in arid regions that is characterized by too high variability in space and time of the amount of precipitation during one rainfall, especially of daily maximums, and also of an annual sum of precipitation for the period in question and its falling in certain seasons of a year.

Analysis of atmospheric precipitation applied various forms of statistical approaches: analysis in time rows, probabilistic, integral-differential and mapping approaches. This enabled study of regularities in precipitation variation and distribution on the basis of time rows.

In plotting maps of precipitation probability – hydrological estimations were done assuming 39% or 63% probability of maximum rainfalls – they used statistical curves of daily distribution of the maximum values in a year (in March and April) as being of the most practical significance for the use of rain runoff.

Analysis of maps of 39% probability of daily maximums of precipitation has shown that their highest value typical of annual maximums is equal to 20-25 mm for a greater part of the territory. Southwards their values tend to increase and reach 20-30 mm, while in the south-east they are more than 30 mm. In March which has a high probability of precipitation, a general diminishment of values of maximums was noted: in the north-eastern area less than 10 mm, in the west 10-15 mm, and in the south-eastern region 20 mm.

A system of maps showing daily maximums of precipitation for most rainy periods is a cartographic model. It enables assessment of water availability in a given region and the use to the maximum of rain runoff, especially in a low rainfall period.

Atmospheric precipitation by virtue of its cycle (i.e. confinement to a certain period in a year and considerable variations of daily rainfalls up to 100 mm) is a reliable source of fresh water and interim, surface runoff in a desert.

16.2 Artificial Watersheds

The main pre-condition for the utilization of atmospheric precipitation is the availability of watersheds. They can be natural with various, impermeable soils and artificial which surface features high water-retaining capacity.

Collection of precipitation from artificial watersheds is practiced mainly in the regions where there are no natural watershed surfaces, and no other fresh water sources.

Although collection of rainfall water from artificial watersheds has its own historical tradition, by the mid-20th century when groundwaters became accessible it had been, in fact, forgotten in most countries of the arid zone. Later on, after development of more effective, water-proof linings an interest in systems of artificial collection of atmospheric precipitation revived again.

One of the main challenges in the construction of artificial watersheds was the development of improved technologies for manufacturing water-proof linings and engineering decisions for enhancement of water yields from natural, watershed surfaces.

Artificial watersheds can be created by various methods: by applying different, water-retaining materials; by creating them directly on a natural surface applying physical, chemical and physico-chemical methods of soil treatment; or with the help of engineering methods that allow increased water yields from natural watersheds; or roofs of houses and shelters having different roofing or water-tight coatings.

Central Asian countries have amassed enormous experience of the technology of collection of precipitation from artificial watersheds. It is of great scientific and practical value and can be used in the establishment of water use systems in other desert regions.

Of special interest at present is a method of artificial treatment of soils. In the last three decades many world countries located in the arid zone have conducted investigations into the creation of watertight and effective linings

In spite of successful implementation of some experimental projects, a method of rainfall collection by creating artificial watersheds has not found wide practical application in any country as yet. The reasons are many but the most important of them is the high cost of chemical reagents, materials and their limited supply.

In piedmont regions of Central Asia water was collected in dry valleys and interim waterstreams; in deserts of Northern Africa and the Middle East in wadi which are dry valleys of temporary or periodically flowing waterstreams; in India, Iran, Egypt and Australia on some slope areas composed of rocks or stony basements of limestone, sandstone, granite.

In Central Asian deserts such watersheds are takyrs. They are typical in the landscape of a sandy desert and represent leveled clay surfaces of different configurations with a gradual slope ranging in size from dozens to hundreds of square kilometers. Takyrs are most often found in the north-western part of Central Asia where their total area is more than 3 million ha. Such surfaces are met also in countries of the Far East and other world regions. In USA such clay watersheds are known as playa, in India as playa-ranna. Ideally they are leveled, inclined and heavily saline vast surfaces nearly completely devoid of vegetation. Watersheds with low water-retaining surfaces are widespread in Israel and are composed largely of light loams.

Development of methods for rational use of atmospheric precipitation was started in 1964 at the experimental center of the Desert Institute of the Turkmen Academy of Sciences which is in a small place called Karrykul in the Central Karakum. A typical takyr of meridional extension was chosen as a watershed. Its length was 2 km, width varying from 0.3 km in its upper part to 1 km in its lower part. According to the results of geodetic surveys, the watershed area was 1.75 km^2. This watershed is longitudinally inclined from the south-east to the north-west, the slope is quite insignificant. In some plots it varied from 0.0006 to 0.0016, making on the average 0.0008. An insignificant transverse slope can be seen.

During copious rainfalls or intensive snow melting, a directed runoff is formed thanks to a longitudinal inclination of the watershed. However when strong winds are blowing (more than 5 m/s) water is easily driven by wind splitting into separate branches because of a transverse slope.

Water flowing from the surface of a takyr watershed after the fall of atmospheric precipitation is slightly enriched with salts and their content usually ranges from 0.2-0.4 g/dm³. Because runoff washes out silty particles from a surface horizon, takyr waters are very turbid and content of suspended matter averages up to 3-4 g/dm³. Such water differs from atmospheric precipitation and, in its essence, is takyr water that provides a basis for a takyr water supply in a desert.

Integrated studies established runoff characteristics: the nature of wetting of a takyr watershed; the variability of precipitation in time and space; the conditions contributing to the formation of a runoff regime, and the method of its estimation.

A specific feature of a regime of takyr waterstreams is the formation of separate episodic floods after copious rainfalls or intensive snow melting. An annual runoff from takyrs is estimated as the sum of floods observed in a year. It is highly variable irrespective of the amount of precipitation in a year. There is no relationship between an annual flow and an annual sum of precipitation.

Engineering and business activities in a desert significantly affect the process of runoff formation and mechanisms of water flowing down from takyr watersheds. This is especially well seen near shepherd settlements, roads, and in the zone of irrigated farming. Due to excessive loads the surfaces of many takyrs are broken; their surfaces are not flat any more, some takyrs are subject to flooding. The intensified processes of overgrowth and sand drifting of a watershed area are responsible for degradation of takyrs as watersheds. Some of them have completely lost their properties as a watershed surface. As a result, in Turkmenistan the area of takyrs participating in runoff formation shrank by more than 30%. And their water collecting properties also deteriorated due to greater initial losses of rainfall and a decrease of effective precipitation.

16.3 The Use of New Technologies

In recent times, in order to increase water yields from natural watersheds even more, new, applied engineering methods using the latest scientific and technological advances have been widely applied. The most promising of them are:

- trapping of filtration flows by cutting trenches to a depth of a screening horizon on a draining surface of the watershed;
- turning a surface of flat watersheds into a corrugated surface with an undulating shape by creating raised corrugations with their follow-on mechanical packing, and smoothing slopes to impart a plastic form to the artificial landscape;
- restoration of a disturbed surface of a watershed by mechanical leveling and returning it to a flat shape; compaction of the surface layer is possible;
- integrated engineering methods in combination with physical and chemical treatment of soil in order to increase water yield from a watershed.

Experience of creation of a corrugated surface of takyrs has shown that this process requires much labor input, and it is necessary to improve its technology and mechanize the work. The most effective method is the creation of an artificial, water drainage system and its periodical clearing of sediments and vegetation. In this case, simple measures will ensure maximum collection of precipitation from a broken surface of the natural watershed.

Therefore, the practice of collection of atmospheric precipitation has shown that efficiency of collection can be attained by increasing water yield of a natural surface or by creating watertight linings. In each particular case, an individual technical decision should be taken with regard to specific local features.

Collection of atmospheric precipitation is by itself only a part of a general problem of their use, and so it is very important to consider local runoff in terms of its storage and management.

Structures for storage and management of collected precipitation together with a watershed or an artificial, watershed site present a simple system in engineering and technical terms. The purpose of storage of atmospheric precipitation is to provide local population and cattle with drinking water for a long period.

Storage systems differ in their purpose and design. Depending on landscape and geological conditions, two methods of storage can be named: overground and subsurface. Water storage requires the following types of facilities: reservoirs, infiltration, catchment and auxiliary.

Overground storage is effected in open and closed structures, the various types of which are given on the enclosed scheme. Closed reservoirs can be of different designs and installed on a ground surface (reservoirs, tanks), and in the ground *sardabas* and *ovdans* (cisterns).

Subsurface storage is effected with the help of fill-in pits *chirle* (wells), infiltration basins, submergible facilities i.e. absorbing wells or infiltration pits in their different combinations too.

In Central Asia, systems for storage of atmospheric precipitation include practically all these types differing only by their design. The most effective method of long-term storage of atmospheric precipitation is subsurface storage; this is called in the research literature a method of creating subtakyr, fresh-water lenses.

Its principle is as follows: during rainfalls, water flowing down from takyrs is directed by embankments to an infiltration submergible system incorporating of a small *oy* type pits or a group of in-fill wells *chirle*. They are usually constructed in the low-lying parts, in places of concentration of water flowing down from a takyr surface. There are normally 5-20 of them, occasionally only 3, but sometimes their number exceeds 100.

In the course of water infiltration through wells, fresh-water lenses are formed at the level of saline groundwaters. Depending on the layout of wells, the submersible systems are subdivided into circular and central. In-fill wells act as operational and water intake wells.

The best results were obtained from simultaneous use of both systems, when from wells in a circular system water was taken for cattle drinking, and from central wells water, being fresher, was taken for the drinking supply of people.

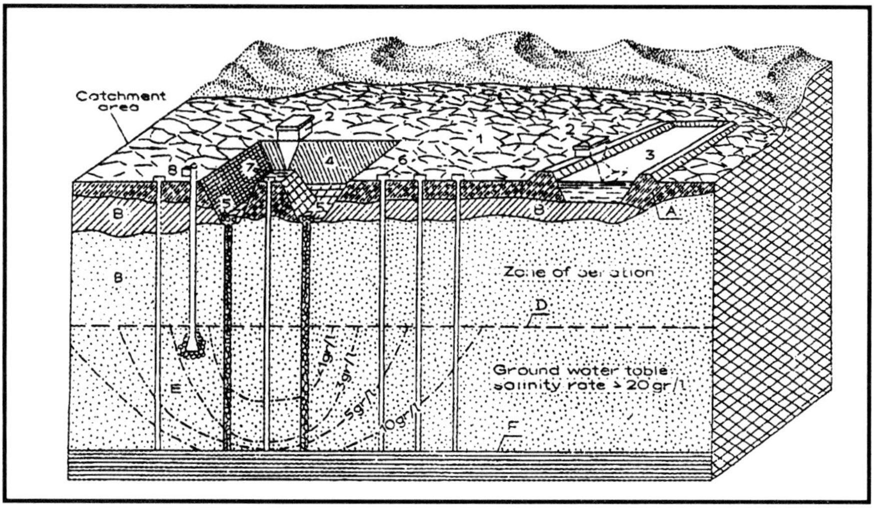

Fig 16.1. Diagram of pilot testing ground for creating underground fresh water lenses in the Karakum **(1)** takyr **(2)** measuring pavilion **(3)** settling basin **(4)** filtration basin **(5)** submerging wells **(6)** observation hole **(7)** cluster of hydrochemical wells **(8)** exploitational well **(A)** clay **(B)** loam **(C)** sand **(D)** Underground water table **(E)** fresh water lens **(F)** water confining stratum

The main difficulty here was to avoid mixing fresh and saline water so as to maintain a fresh-water lens for a longer time.

With the advance in science and technology, a principle of construction of overground and subsurface facilities for storage of precipitation has changed. Thus, *kakis* appeared that had been constructed by excavation machinery. They created significant reservoirs capable of storing 100,000 m^3 of water and even more. Designs of closed facilities have also altered significantly. Dug pits were replaced by engineering wells with armoring lining. The latter completely lost their functions of submersible structures, and were intended only for water intake with the help of strip water lifts.

Therefore, methods and techniques of people's hydraulic structures tested over centuries were completely replaced by engineering structures, the facilities for collection and storage of precipitation were also reconstructed. In spite of such progress in methods of accumulation and storage, the water reserves thus created were limited, and in some cases failed to meet the growing demand.

Creation of new, more efficient systems should be based on novel technologies and technical designs that will ensure guaranteed water supply in a desert. The Desert Institute conducted researches along this line during many years and established in the Central Karakum a unique, experimental base for:

- development of engineering designs for collection of an adequate amount of precipitation and efficient technologies for their storage;

- establishment of an experimental hydrocomplex on the basis of rational management of precipitation.

This experimental base was used for development of new designs for subsurface storage of atmospheric precipitation as one of the most promising ways for its rational management. Surface runoff from takyrs entered a settling basin, and then through an infiltration system dropped into subsurface collectors in a saline water aquifer where an artificial lens of fresh groundwaters was formed (ALFUW). A diagram of subsurface storage of precipitation is shown (Fig. 16.1.).

The efficiency of this system completely depended on specific natural features of a region and water quality. Investigations for ALFUWs were carried out in very difficult conditions which arose from a complicated structure of the zone of aeration 15 m thick, a high groundwater salinity (20-30 g/dm^3), and high water turbidity (3-4 g/m^3). Besides, a faulty choice of operational parameters of an infiltration structure 5,000 m in capacity which had been designed for a single filling, failed to control the optimal volume of runoff from the takyr that was more than 100,000 m^3 during one flood. In spite of this they proved that it is possible to construct ALFUWs in 3-4 years, and its storage capacity will be enough to provide water for up to two flocks of sheep during a year.

Increase of operational parameters of the structure allowed a three-fold increase of capacity for a single filling. An improved technology of plunging surface waters by construction of absorbing wells of large diameter, and the building of pits which intensified infiltration with insignificant, preliminary clarification of plunged water in a settlement basin contributed to a 10 times higher efficiency of this infiltration system. It was also noted that ALFUWs acquired a more compact shape with better dimensions for operation.

16.4 The Optimal Systems of Precipitation Use

Analysis of the results of experiments with submersible systems of different designs has shown that the most efficient choice ensuring intensive plunging of water in conditions of a complex structure of the whole zone of aeration is a system with absorbing wells of a large diameter. Under complicated, hydrogeological conditions in the upper part of the zone of aeration, which is usually penetrated by an infiltration basin to a depth of 2-3 m, no less effective is an infiltration structure in combination with pits which intensify infiltration. Construction of ALFUW by a basin method gives good results only in regions with favorable filtration conditions in the zone of aeration.

Introduction of a hydro-infiltration system allowed the development of the principles of zoning of a desert territory in terms of conditions for collection and storage of atmospheric precipitation.

Design of a system for subsurface storage of surface waters includes construction of an artificial watershed site lined with a film or other water-tight material.

Such a system of atmospheric precipitation storage is, in fact, a water supply hydrocomplex in a desert. Creation of fresh water reserves by collecting atmospheric precipitation from natural and artificial watersheds open a possibility to organize small-oasis farms. Experience obtained in the Karrykul hydrocomplex in the Central Karakum can be used in organization of water supply of pastures to meet requirements in drinking water of people and cattle.

17. The Technology of Making Lenses of Fresh Underground Water in Deserts

I.D. Mamieva

ABSTRACT: Artificial Lenses of Fresh Underground Water (ALFUW) are manmade reservoirs. Their successful functioning, apart from local geomorphologic features, depends on a complex system of infiltration and exchange between fresh and saltwater. This is because the freshwater created are surrounded by water with increased mineralization. The Desert Institute has much experience with research in this field, and has analyzed the process during the various cycles of which exact details are given. With the creation of ALFUWS, sheep breeding in particular can be furthered.

The organizational problem of any kind of water supply will always be of great importance. The significance of this problem for Turkmenistan is explained by the increasing tendency to deterioration of surface water quality and the Aral drying up. So the relevance of elaboration of the technology of rational utilization of atmospheric precipitation, forming temporary surface drains on takyr soils, is evident enough. The collection and storage of temporary surface water possesses a number of advantages, of which the main are:

- the possibility of making sufficient water reserves, providing a guaranteed water-supply in case of years with little water;
- the supply of high quality, utilizable water;
- the simplicity of the exploitation of the system of storage/magasination.

As a result of experiments in artificial lens forming of fresh underground water with mineralization of more than 17 g/l in Karrykul (Central Karakum) using atmospheric precipitation during 4 years, the lens has been formed, restricted by the mineralization content up to 10 g/l, to a volume of 20,000 m^3 and with the maximal size about 7 m. Under the water-output coefficients, varying between 0.11-0.20, it is possible to take up to 4,000 m^3 of water of the indicated mineralization out of this lens.

17.1 Factors Determining Lens Utilization

The utilization of lens water for various economic purposes depends on the extent of their mineralization and chemical composition. They, in their turn, are determined by geochemical processes, occurring under the transformation of precipitation into underground water in the system: the source which is precipitation, then water collection followed by the infiltration system into the zone of aeration-subsoil water; that is mineralization and chemical composition of artificial lenses of fresh underground water (ALFUW) depends on the quality of takyr water, the extent and type of salinization of soils of the zone of aeration, the quality of underground water of the main horizon and on others.

The analysis of observation over 25 years permitted the main geochemical processes in the system and formation of ALFUW to be studied.

17.1.1 Geochemical Factors

The atmospheric precipitation, forming the drain on the takyr, is a source of feeding for lenses. Its magnitude should be not less than 16 mm during one period of heavy rain when the ground is dry, and not less than 10 mm when the ground has already been wetted (14-20%). The mineralization of water of liquid atmospheric precipitation is 0.1-0.2 g/l, snow – 0.3-0.4 g/l. They are, as a rule, calcium or natrium hydrocarbonate in the chemical composition of water.

The higher the level of easily soluble salts which are dissolved from the upper 10 cm takyr stratum, the more they determine the chemical composition and extent of the mineralization. Two types may be differentiated in the takyr Karrykul according to the extent and character of the salinity:

- the natural takyr, being the main area of activity of the water collection with the extent of salinity 1.1-3.7% and natrium chloride-sulphate type of salinization;
- the takyr, exposed to overgrowth with grass as a result of the anthropogenic influence with the extent of salinity 0.4-0.7% with the calcium sulfate type of salinization.

In the processes of drain forming, the water of atmospheric precipitation are enriched by ions of chlorine and natrium as a result of which their mineralization increases up to 0.6 g/l. The contents of the ions of chlorine and natrium in water increase by 2-5 times. The change of mineralization is connected with the regime of the drain forming, the time of the contract of water with the underlining surface and evaporation.

17.1.2
The Infiltration Process

Under the entrance of a drain from the takyr into the infiltration foundation pit, intensive water infiltration is observed. The evaporation from the water surface compared with the infiltration is not significant. In connection with this, the mineralization and chemical composition of infiltrational water in respect to the takyr drain do not vary. During the prolonged work of construction, its floor and sides are colmatated, which sufficiently influences the regime and speed of infiltration. Under it the coefficient of the infiltration varies from 0.76-0.5 m/day. The thickness of the crust, formed during one filtercycle, makes up to 2 cm., and that of the subcrust silted layer 5-12 cm. Given the colmatation of the bottom of the foundation pit, the quantity of evaporation and infiltration become measurable by the same standard. All this results in the increase of the infiltration water up to 0.8 g/l. The type of water remains as before.

The main changes of the quality of water are observed in the aeration zone, where the surface drain is transformed into the underground one and the main migration of the water's chemical elements occurs. In the area of the formation of the ALFUW, the aeration zone consists of fine-grained sands with streaks of clays and loams to a depth of 12 m, under which the quartz feldspar sands of the Garagumyskaya to a further depth of 15 m are revealed.

The salinity of a profile is 0.8-2.49% in the upper part, and 0.14-0.04% for 100 g of soil in the lower thickness. The initial salinization is natrium chloride/sulphate. The capacity of absorption varies within the limits from 1.83 for the sands of the Garagumyskaya up to 0.6 mg-eqv, and up to for 100 g for the fine-grained sands. The clayey streaks have not been sustained according to the spreading of the stratum and have, as a rule, a lens-shaped form. The capacity of absorption of loams and clays is 12.0-14.9 mg-eqv.

In the absorbing complex of clays, cations of the alkaline soil metals prevail. All the kinds of aeration have been gypsummed to a depth of up to 15 m. According to Kravtsov's data in contour, limited by sizes of the surface of the foundation pit (50 x 50) and the depth of aeration zone from the bottom of the foundation pit up to the level of subsoil water (12 m), the summary quantity of hypothetical salts, being contained in lithologic differences, made up 205 t.

The subsoil water with a small slope of the flood mirror 0.002-0.0003 on C3, have the initial mineralization 17-22 g/l and natrium chloride composition.

17.2
The Formation of the Lens

During the period of ALFUW forming about 30,000 m^3 of water were poured into the foundation pit, approximately 25,000 m^3 have been filtered. The evaporation has made up the rest of the amount.

Under the first filling of the foundation pit (water volume 5,100 m^3) the lens has not been formed yet, because the water has saturated the aeration zone, where the salts are significantly distributed; the output of salts then flows into the saturation zone.

In the upper 7 m thick stratum, the contents of salts decreased, but from the depth of 7 m and up to the level of the subsoil water increased by 0.4-1.0%. Under it the contents of the ions of HCO_3 markedly decreased, and then remained approximately constant. At the expense of gypsum hydrolysis, contained in the subsoils of the aeration zone, the contents of the ions of Ca^{2+} in water increased.

Under the following movement of water into the foundation pit, washing out of the easy soluble salts from the aeration zone and the formation of the ALFUW takes place.

The contents of the easily soluble salts in the subsoils are determined by the characteristic formative processes of the accumulation of salts. The character of the salinity of the subsoils reflects the quantitative contents of salts according to the horizons of the lithologic profile and their chemical composition.

The main feature of the investigated subsoils is the considerable content of gypsum (up to 7%); in principle this does not result in salinization, but instead decreases the filtration properties of the subsoils.

17.3 Salts Redistribution

The whole process of the redistribution and the washing out of salts of the aeration zone may be divided into 3 stages:

- desalinization of the subsoils of the aeration zone, when 80-90% of easily soluble salts are washed out;
- desalinization of subsoils at the expense of lysis of the intrasoil gypsum;
- there is a change of ionic composition caused by the water solution entering into a reaction with a rock strata through its pores.

The first stage ends after the maximal drop of the filtrate mineralization, and reflects the displacement of solutions out of gravitational and capillary pores larger than 5-6 λm.

The second stage is characterized by the increase of the ions of Ca^{2+} and SO_4^{2-} in water at the expense of the gypsum hydrolysis.

The third stage is characterized by the exchange reactions of the ions solution-stream and the absorbing complex of clays, loams and loamy sands with a large capacity of the exchange.

Under it, the major part of Ca^{2+} ions released at the expense of desalinization of the intrasoil gypsum is consumed through absorption by soil colloids and displaces the ions of Na^+ and Mg^{2+}. At a particular point, equilibrium is established between dissolved gypsum and absorbed sodium. The sulfate ions are accumulated in the solution, and they inhibit the dissociation of the gypsum molecules.

Table 17.1. Distribution of the Hypothetical Salt composition (mg eqv) in the aeration zone at the site of the lens formation

Layer thickness	$Ca(HCO_3)_2$	$CaSO_4$	$Mg(HCO_3)_2$	$MgSO_4$	$MgCl_2$	$NaHCO_3$	Na_2SO_4	NaCl
Default salinity								
0-1 m	0,0324	0,0544		0,0481			0,2201	0,4271
0-5 m	0,0324	0,1395		0,0361			0,3520	0,6270
0-10 m	0,0364	0,0850		0,0240			0,2690	0,4320
0-15 m	0,0360	0,0850	0,0170	0,0230	0,0086	0,0084	0,4750	0,3630
% of easily soluble salts				2,6	1,0	1,0	54,1	41,3
I filtercycle (W=5.000 m^3 water)								
0-1	0,0284	0,2108		0,0403			0,6610	0,4095
0-5	0,0284	0,2108	0,0365	0,0403		0,2873	0,4080	0,4095
0-10	0,0500	0,4320	0,0210	0,0494		0,2873	0,5440	0,5210
0-15	0,0460	0,2755	0,0210	0,0337		0,2873	0,4140	0,4075
% of easily soluble salts				3,1		35,1	26,2	35,6
II filtercycle (W=10.000 m^3 water)								
0-1	0,0334	0,1168		0,0276			0,1508	0,0175
0-5	0,0334	0,0334		0,0276			0,1508	0,0175
0-10	0,0317	0,0773		0,0207			0,0884	0,0167
0-15	0,0294	0,0494		0,0141			0,0672	0,0131
% of easily soluble salts				14,6			71,5	13,9

With further filtration the products of exchange reactions are removed out of rocks, and that increases the gypsum solubility up to 5.6 g/l.

The calculation of the hypothetical composition in the aeration zone before the filling showed that the main salts are the easily soluble salts (NaCl, Na_2SO_4, $Mg(HCO_3)_2$, $MgSO_4$), the weakly soluble ones – $CaSO_4$, Ca $(HCO_3)_2$. Within the whole 15 m thick stratum, there are 85 to 87% of easily soluble salts, out of which NaCl and Na_2SO_4 account for 41.3 and 54.1%, respectively (Table 17.1.).

After the first water intake, the following salt redistribution has occurred: the natrium salt contents in the horizon at 4-5 m has decreased by 0.8-1.0%. In the horizon at 5-6 m, the decrease of salt of natrium is not significant, but in the horizon at 11-15 m the contents of NaCl increased from 0.02 to 0.19%. Thus, the sodium salts as mobile compounds have been leached into the lower lying horizons of the aeration and partially into the zone of saturation. The store of easily soluble salts after the first water intake has been reduced by more than 10% and has made up 77.1%, out of which NaCl contributed 35.6% and Na_2SO_4 – 36.2%. After the second water intake into the infiltration basin (10,000 m^3 of water capacity) the salinity of the subsoils of the aeration zone decreased by 2-3 times with respect to the initial one. Under it further easily soluble salts were washed away so that their contents made up 54.5% in relation to the initial one.

As a result of two intakes, 89% of easily soluble salts have been removed from the whole 15 m thick strata.

The change of the salinity of the subsoils of the aeration zone and the chemical composition of the solution stream correlates well with the variations of the chemical composition of the subsoil water of the lens (Table 17.2.).

17.4 Geochemical Stages of Lens Formation/Supply

With respect to prevailing geochemical processes in the forming and supply of the lens, 3 stages may be separated:

- The aeration zone flushing to evacuate easily soluble salts, their partial removal into the saturation zone and the formation of the zone of the intensive water exchange in upper horizons of the underground water beneath the infiltration constructions. Under it, the displacement of soluble substances occurs mainly under the influence of the convective transfer but also to some extent through processes of the exchange of ions, salt precipitation, and the formation of ionic couples and complex compounds in the solution.
- The forming of the chemical composition of water of the lens, when substantial volumes of fresh water come into the saturation zone and the "plunger-type" displacement of saline water occurs, where the main role is played by convection.

Table 17.2. Distribution of the Hypothetical Salt composition (mg eqv) in the Saturation Zone at the Site of the Lens Formation (Karrykul, pit 39)

Sampling depth	$Ca(HCO_3)_2$	$CaSO_4$	$MgSO_4$	$MgCl_2$	$NaHCO_3$	Na_2SO_4	NaCl
Initial salinization							
15 m	0,0324	0,01734	0,3756	-	-	0,0913	1,3309
16 m	0,0324	0,01734	0,3756	-	-	0,0913	1,3309
17 m	0,0324	0,01734	0,3756	-	-	0,0913	1,3309
I filtercycle (W=5.000 m^3 water)							
15 m	0,0243	0,1598	0,3696	-	-	0,9301	0,5534
16 m	0,0227	0,1578	0,4363	0,0067	-	-	1,2788
17 m	0,0259	0,1482	0,3848	-	-	0,0562	1,2753
II filtercycle (W=10.000 m^3 water)							
15 m	0,0267	0,1356	0,3074	-	-	0,9359	0,5383
16 m	0,0243	0,1496	0,3089	-	-	0,9656	0,5497
17 m	0,0275	0,1503	0,3089	-	-	0,9507	0,5405
Supply absence							
15 m	0,0202	0,0150	0,0378	-	-	0,2762	0,5097
16 m	0,0081	0,0408	0,0480	-	-	0,5708	0,6134
17 m	0,0150	0,0493	0,0859	-	-	0,5001	0,6046
III filtercycle (W=5.000 m^3 water)							
15 m	0,0059	0,0707	0,0481	-	-	0,2404	0,0760
16 m	0,0065	0,0966	0,1091	-	-	0,4831	0,5148
17 m	0,0178	0,0326	0,0084	-	-	0,0845	0,4917
IV filtercycle (W=5.000 m^3 water)							
17 m	0,0130	0,0326	0,0268	-	-	0,0170	0,2776
16 m	0,0065	0,0694	0,0448	0,0033	-	-	0,3299
18 m	0,0081	0,0748	0,0420	0,0119	-	-	0,3591
% of initial quantity	75,0	43,1	88,8				73,0

- The period of absence of the supply to the lens. The main process is the molecular diffusion i.e. the process of leveling of groundwater concentrations, temperature or pressure.
- The lens supply; the chemical composition of water under it remains former, but increases the freshened lens capacity.

As a result of the first filtercycle (5,000 m^3), the initial mineralization of the removal out of the aeration zone increased by 2-5 g/l. With the prolonged feeding of the lens (20.000-25.000 m^3 in volume), soluble salts are washed away out of the aeration zone subsoils, the mineralization of the underground water decreases up to 2 g/l, and the definite lens composition is formed. The lenses in the chemical composition of water are the sodium/calcium sulfate/chloride type.

Given a lack of water supply, the diffusion processes are manifested in the leveling of the concentration and the formation of the transitional zone from the fresh water to the salt ones. Given 18 years of absence of supply, the mineralization of the lens water increased up to 4.5 g/l.

Under the lens supply with a water volume up to 14.000 m^3, the mineralization decreased up to 2 g/l, the type of water did not change in the chemical composition.

The experiment showed, that the lens of fresh water having been formed, and not receiving any input, preserves its total volume, decreases the maximal depth, while increasing the width. The lens volume, restricted by the contour of the given mineralization, under other equal conditions of its artificial forming will have more or less the initial mineralization of the underground water.

17.5 The Results of the Experiment

Thus, the experimental works in magazining of a drain by the basin (foundation pit) method show, that it is necessary to sink about 30.000 m^3 of fresh water into the aeration zone for the forming of fresh lens, suitable to the exploitation. Given the average, annual drain of 10-15 thousand m^3 out of km^2 such a lens may be created for 3-4 years. In subsequent years the lens may be exploited and supplied.

The main feature of the exploitation of lenses of fresh underground water compared with the exploitation of the usual, water-bearing horizons are that the fresh lens is surrounded by the water of the increased mineralization. Therefore during the exploitation, together with fresh water the mineralized ones, which at a certain moment break out into the fresh water-taking, which gives way to their motion.

The mineralization of the pumped out water increases uninterruptedly and becomes unfit for use. It is important to note, that under the influence of the exploitation the dome of salt water under the duct filter is formed independently of the volume of non-salty (fresh) water; the type of duct and the time of pumping out play the main role here, but whether early or late the seeping through of salt water to the filter will certainly take place.

17.6 Differentiation of Lenses

The artificial lenses of fresh, underground water differ from the large, natural, subsandy lenses:

- in genesis and conditions of forming, hence, for the conditions of the feeding;
- in the low volume of the fresh water zone;
- in the low volume of the exchange zone from fresh water to the mineralized ones. The ALFUWs are formed under the influence of anthropogenic activities. Through them the fresh water fully replace the salt water, and the transitional zone nearly does not have the time to be formed, and makes up from one to several meters.

In principle, the calculation of water pumping out of the fresh lenses comes down to the choice between the optimal exploitation regimes: either exclude the seeping through of salt water or, with their seeping through, then the mineralization level of the pumped out water must not exceed tolerable levels.

During the choice of the ALFUW exploitation regimes, it is necessary to take into account the enumerated features. The low volume of the fresh lens and big concentrations of mineralization in the exchange zone from fresh to salt water do not permit using the same method of the calculation of the water-taking of large, natural lenses for the ALFUWs.

The experiment shows, that during the exploitation of the little, artificial lenses the water salinity usually increased even for pumping out of small amounts.

17.7 Estimates from the Modeling

The mathematical modeling of exploitation gives the following estimates: the maximal volume of a single duct, of the ALFUWs, under the possible mineralization of 1 g/l during 100 days (summer) is 30 m^3/day; with the duct taking in 320 m^3/day, fresh water may be obtained only during 2.5 days; with the duct allowing 80.4 m^3/day – 16.5 days; with duct set for 40.2 m^3/day – 40 days. The water volumes, obtained by these different settings are 804.25 m^3, 1326.6 m^3 and 1608 m^3, respectively (Fig. 17.1.).

During the process of increasing the pumping out of water, mineralized water intensively flows in the first 10-20 days, and then tends towards stabilization (Fig. 17.2.).

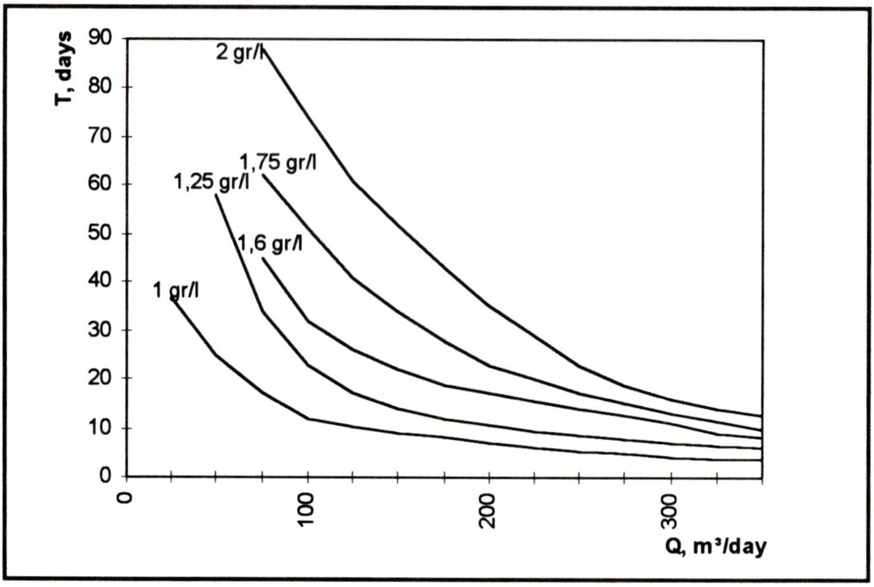

Fig. 17.1. Duration of pumping out (T) water with a miniralization of (C, g/l) depending on debits (Q)

The Karrykul lens exploitation requires the elaboration of non-standard technical means, oriented upon few ducts and insignificant depths of selection.

The information obtained in the course of the complex experiment of the lens creation, makes up the data base, necessary for the identification of the ALFUW models.

The estimation of the possibilities of lens water for various economic purposes showed, that with the mineralization up to 2 g/l their utilization is possible in small-oasis agriculture for light soils with annual washings out; water with mineralization of more than 2 g/l are practically unfit for watering.

Lens water with mineralization up to 5-6 g/l may be utilized for animals. The chemical composition of the lens formed corresponds to the possible norms for the animal ponds in conditions of remote cattle and flock breeding.

17.8 The Possible Impacts on Farming

In the Central Karakum, with the radius of flock movement 7-8 km and the norm of pastures 9 hectares per sheep, up to 1,700 head of small cattle may be kept in the area of the pond point. About 3,000 m^3 of fresh and poorly mineralized water a year are required for the pond and the water supply of the service personnel.

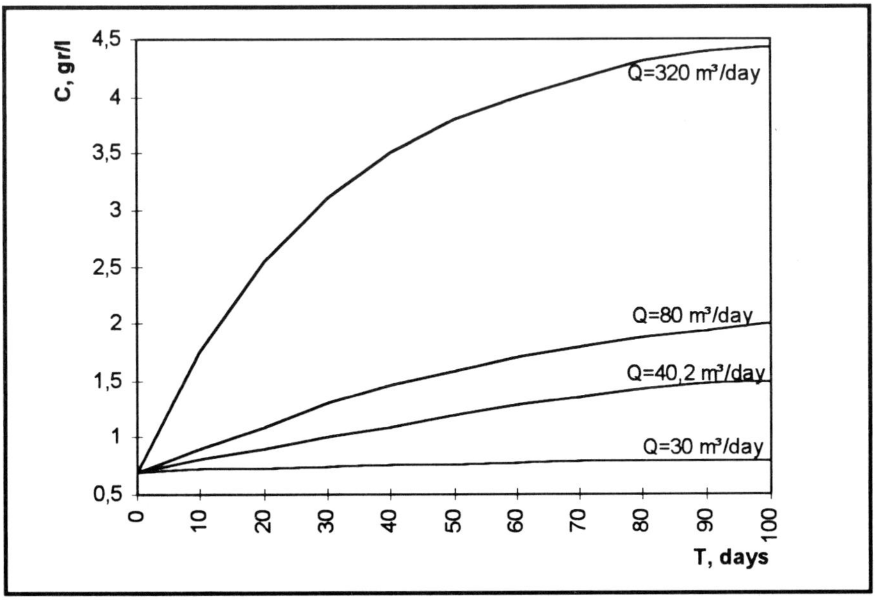

Fig. 17.2. Pumped out water miniralization change (C) under various debits (Q)

Contrasting required water volumes in the pond with the created reserves for 4 years under the existing lenses and the forming of new ones of fresh water, one should note the considerable amount of water, which may be utilized for the water supply of the desert pastures, the creation of a small oasis economy, the provision of other small consumers and for the case of years with little water.

18. The Use of Drainage Water

M.V. Kolodin, G.I. Rabochev

ABSTRACT: After detailing water statistics for Central Asia and the problems of increasing usage, the question of drainage water recycling is addressed. Central to this is the extent and kind of salinity. The use of the sodium adsorption ratio is explained. The results of experimental schemes in Central Asia including the affects on plant yields are listed along with environmental dangers and antidotes. Details of a new technology developed by the Desert Institute are given and contrasted with a similar scheme in Arizona. Finally the cost estimates for a pilot project are calculated.

Fig. 18.1. shows distribution of water resources among Central Asian countries. The total resources of surface and ground water in the five CIS Republics are estimated at 229.8 km³/year. Water is the most vital, natural resource. The diagram shows that the most acute, water deficit is observed in Uzbekistan and Turkmenistan where the available water resources are, in fact, exhausted.

Beginning in the 1960s in Central Asia, intensive construction of hydraulic structure started. If at that time the area of irrigated lands in this region was 5.3 million ha, then by 1994 it had already reached 10.2 million ha. With the development of irrigated farming the volume of drainage water (DW) was also increasing.

18.1 Water Statistics for Central Asia

Earlier irrigated lands were mostly found in upstream areas of rivers where a kind of "dry drainage" was in action. In recent years, however, mostly flat lands in the middle and lower reaches of rivers with a high drainage coefficient are under irrigation, and this increased drainage flow. Thus, in Turkmenistan in 1960 the area of irrigated lands was 496,000 ha, and the drainage flow from them was 0.1 km³. By 1994 the irrigated area had increased to 1,728,000 ha or by 350%, while the drainage flow had increased to 6.6 km³ or by 6,600%.

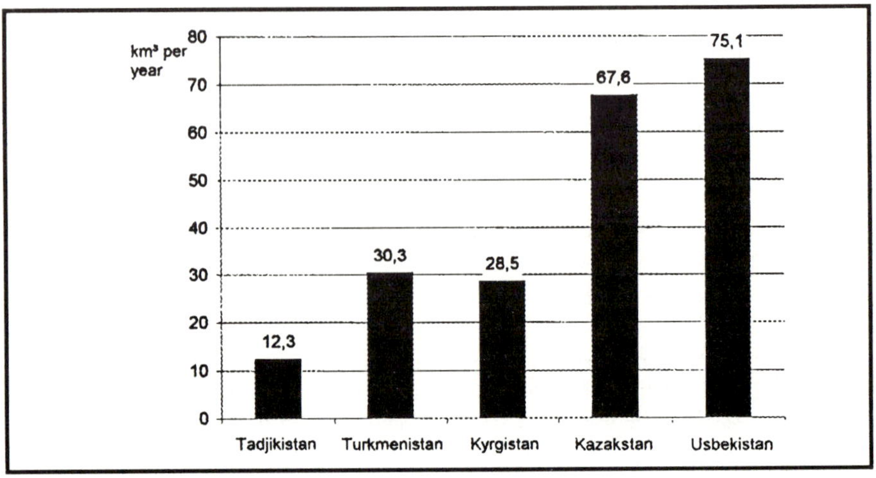

Figure 18.1. Distribution of water resources in Central Asian countries

Different authors provide different figures of the total drainage flow in the Aral Sea basin: the estimates range from 33.1 (Sanin *et. al.* 1991) to 45.8 km^3/year (Glazovski 1990), while resources of drainage water might vary from 47.9-52.2 km^3 a year (Razakov 1991). In general, drainage water resources in this region are 59.1 km^3 a year, of which 1.1 km^3 consists of effluents and nearly 48 km^3 are drainage water.

Increased volumes of drainage flow can be explained by a low farming culture and a low engineering level of an irrigation and drainage system which consists mostly of earthen canals, thus, great water losses to seepage. On the other hand, in this region the actual use of water for irrigation (gross) varies from 12,000 m^3/ha in Kyrgyzstan, 15-16,000 m^3/ha in Kazakstan and Uzbekistan and 18-20,000 m^3/ha in Tadjikistan and Turkmenistan. Glazovsky estimated that in 1980 for irrigation of 7.9 million ha the excess of irrigation water in comparison to the biological requirement was 44 km^3. By 1994 the area of irrigated lands had increased to 10.2 million ha, and excessive use of water increased.

Since the 1970s in Turkmenistan, a growth of water consumption has been observed. Every year's water intake increases by more than 1 km^3. In 1970 water consumption in the state was 12.92 km^3 a year, then by 1995 it had risen already to 27.61 km^3 a year. At the same time drainage flow increased from 2.16 km^3 a year in 1970 to 6.1 km^3 in 1990 and 4.8 km^3/year in 1995. Diminishment of a drainage flow in 1995 can be explained by taking special technical actions to reduce the drainage flow.

18.2 Using Drainage Water

Of the total amount of discharge water a major share of drainage water (more than 26 km^3 with a salinity up to 5 g/l) can be used for irrigation of forage and other crops. Mediterranean countries, USA, India, Pakistan and Australia have accumulated rich experience in the use of marine, saline groundwater and drainage water for irrigation of different crops.

The possibilities of using drainage water for irrigation depends on many factors and conditions: soil texture and soil salinity, salinity of drainage water and a salt composition in them, salt resistance of agricultural crops and feasibility indices of irrigation systems. Drainage water can be best used for irrigation of sands and coarse-textured soils featuring high filtration properties and a low absorbing capacity.

Using drainage water for leaching, regular or periodic irrigation is dependent on a provision of drainage systems on a territory. For regular irrigation, soil sites with good drainage characteristics of a soil series are chosen. If the content of chlorine-ion in drainage water is up to 0.3 g/l, then the requirement in drainage increases by 10-40% in comparison to plots irrigated with fresh water.

For creation of an optimal salt regime in a root zone, the soils in areas of regular and periodic irrigation with drainage water should be either non-saline or slightly saline. It is not recommended to irrigate medium and heavily saline lands with drainage water.

The economic advisability of application of drainage water for regular irrigation can be determined by the efficiency of capital costs i.e. a comparison of a net income increment with costs of irrigation with drainage water in comparison to fresh water. The economic advisability of application of the drainage water for periodic irrigation is factored by comparison of additional costs with additional, economic benefits obtained by overcoming of an irrigation water deficit.

18.3 Water Salinity and Salt Composition

No less important are drainage water salinity and salt composition. These indices determine the irrigation properties of water. In order to avoid negative implications for soils and diminishment of fertility under irrigation with drainage water, their quality is assessed by different criteria taking into consideration a risk of total and chlorine salinization, alkalisation and magnesium salinization etc.

Calcium sulfate, magnesium and calcium carbonates refer to salts that are not harmful for plants. Harmful toxic salts include all chlorides, sulfates of sodium and magnesium, as well as soda. The amount of these salts and their ratio determine – applying different coefficients – irrigation properties of drainage water.

18.4 Sodium-Adsorption Ratios

In classifications, sodium-adsorption ratios (SAR) characterizing a risk of alkalisation are often used:

$$SAR = Na^{+} : \sqrt{0.5\left(Ca^{2} + Mg^{2}\right)}$$

where cations are expressed in meq/l.

Table 18.1. gives qualitative appraisals and conditions of drainage water application for irrigation. At SAR < 10 water does not have a risk of alkalisation, at SAR < 11-15 the risk grows, especially for loams and can be used on soils with good, water permeability properties after dilution by fresh water.

At SAR > 16 a risk of alkalisation for all types of soils is very high. Such drainage water can be applied for irrigation only after dilution with fresh water. If the content of chlorine ions in drainage water is more than 0.3 g/l, these water are harmful for plants.

18.5 Experimental Irrigation

In CIS countries, especially in Central Asian countries, for many years on trial fields and under full-scale conditions, they conducted experimental irrigation of agricultural crops with drainage water. In Kyrgyzstan (Kenes-Anarkha) under irrigation with drainage water (4-5 g/l), the yield of barley was 80-90 quin/ha, corn (kernel) 35-40 quin/ha (1 quintal=100 kg).

In Uzbekistan in the Golodnaya steppe, they conducted an experiment over 8 years on the irrigation of cotton with drainage water from the collector "*Shuruzyak*" with a water salinity level 2.84 g/l, and from vertical drainage wells with a salt level 4.44 g/l. An average yield under irrigation with river water was 32.7 quin/ha; under irrigation with water from a collector with a chlorine level 0.45 g/l the yield sank by 3.1 quin/ha or 9.5%; under irrigation with water from wells with a chlorine level 1.22 g/l the yield dropped by 6 quin/ha or 18.3%. At a chlorine level of more than 0.5 g/l, the water becomes toxic and inhibits plants.

In Turkmenistan on the Kopetdag piedmont plain, on takyr-like, serozem, low-saline soils (of medium and coarse texture) under irrigation with drainage water (3.9-4.5 g/l) the yield of green mass of the following forage crops was high (corn: 444 quin/ha; jougara: 720 quin/ha; Sudan grass: 700 quin/ha and sunflower: 796 quin/ha, which is only 5-10% lower than under irrigation with fresh water.

Favorable conditions for the use of drainage water for irrigation are found in the Lebapsky velajate where there are large areas of well-drained, sandy soils, and significant resources of drainage water low in salts.

Table 18.1. Qualitative assessment and conditions of drainage water application for irrigation

Water quality	SAR	Salt content in drainage water	Sum of toxic salts		Dilution with fresh water
Very good	5	0.5-1.0 g/l	0.1 g/l	10%	0%
Good	6-10	1.0-2.0 g/l	0.4 g/l	20%	0%
Average	11-15	2.0-5.0 g/l	1.8 g/l	35%	40-50%
Fairly unsatisfactory	16-20	5.0-8.0 g/l	4.0 g/l	50%	60-70%
Very unsatisfactory	> 20	> 0.8 g/l	> 4	> 50%	> 70%

In the Lebapsky velajate, irrigation of sandy soils with drainage water (salinity 2.1-2.5 g/l) from the main, left-bank collector was carried out on an experimental field of 3 ha belonging to the Chardjou station of the Desert Institute. Later under full-scale conditions they irrigated 2,000 ha on the Sakarsky poultry farm. At an application rate of 8250-9550 m^3/ha the yield of green mass of corn was 440 quin/ha and of Sudan grass 702 quin/ha. A unit cost of one ton of green mass was US$ 6.8-8.0, which is less costly than under irrigation with fresh water, while economic efficiency of irrigation of one hectare with drainage water in comparison to river water was US$ 50-60.

18.6 Possible Environmental Dangers

Drainage water are harmful for the natural environment and lead to a deterioration of the ecological situation in all irrigated regions. After disposal of drainage water into water bodies, rivers and irrigation sources the water quality in them sharply deteriorates. Diversion of drainage water along waterways to natural depressions or directly into a desert causes groundwater rise and salinization of lands and pastures.

Apart from mineral salts, drainage water also contain chemical fertilizers, agricultural chemicals, petroleum products and other harmful substances that contribute to pollution of water, soil and foodstuffs. If drainage water cannot be directly used for irrigation, then in order to improve an ecological situation the other, more polluted part of them should be subject to decontamination, treatment or desalinization. There are three possible technical decisions of water protection measures:

- remove to streaming pond (Option I);
- simple freshener (Option II);
- complex demineralization (Option III).

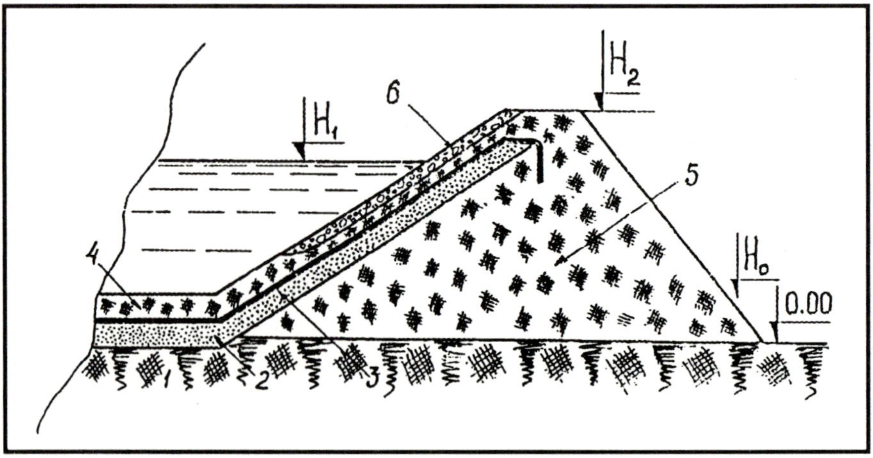

Figure 18.2. Evaporation pond of an engineering type **(1)** a pond bed **(2)** packed ground **(3)** polyethylene film **(4)** a protection layer covering a film **(5)** embankment **(6)** reinforcement of an embankment crest

18.7
Antidotes to Pollution

Highly saline and polluted drainage water should be diverted to special accumulation reservoirs or evaporation ponds. In the arid zone, from 1,500-2,000 mm of water is evaporated during a year, and decontamination of drainage water in evaporation ponds will be most efficient.

To avoid pollution of groundwater, while selecting a site for construction of ponds, it is necessary to take into consideration hydrogeological conditions on this site. It is better to construct ponds on low permeable or impermeable lands. If fissured or karst rocks are developed and they are not overlain by a water-impermeable layer, then it is necessary to conduct special proofing measures to prevent filtration of drainage water through a pond bed and embankments as is shown in Fig. 18.2.

Table 18.2. gives technical economic indices and energy use for decontamination of drainage water in evaporation ponds in the arid zone of Central Asian republics (option I).

Other countries practice widely used hydrobotanical (bioplateau) or sorbtion methods to improve the quality of organic treatment of waste water. As is known, higher aqaueous plants (hyacinth, reed, cane, cattail and others) are capable of absorbing giogenes and mineral substances from water; mineral substances are not absorbed so well. Experiments have shown that in some CIS countries a bio-plateau is not effective for treatment of drainage water with a salinity of 2-4 g/l,

Table 18.2. Enlarged costs of decontamination and desalinization of 1 km³ of drainage water on plants with a capacity of 1,000 m³/h

Technology of drainage of water treatment	Maximum salinity	Economic indices [million US$]		Energy use	
		capital costs	operational costs	electric capacity [MW]	conventional fuel [million t]
Decontamination (I)					
in evaporation ponds	5 g/l	4100	210	57	0.158
	10 g/l	4330	225	57	0.158
	20 g/l	4540	240	57	0.158
	35 g/l	5430	270	57	0.158
Traditional desalinization (II)					
electrodialysis	3.5 g/l	1080	460	314	0.86
	5,0 g/l	1600	700	440	1.29
	10,0 g/l	2140	1120	500	2.58
reverse osmosis	3.5 g/l	1200	590	235	0.66
	5,0 g/l	1320	690	346	0.96
	10,0 g/l	1560	1020	691	1.92
Complex demineralization (III)					
distillation	5,0 g/l	1800	785	329	20.1
	10,0 g/l	2280	780	329	20.1
	20,0 g/l	3220	770	329	20.1
	35,0 g/l	4620	760	329	20.1
electrodialysis	3.5 g/l	1610	240	506	6.0
	5,0 g/l	2110	450	608	6.2
	10,0 g/l	2800	780	1157	7.8
reverse osmosis	3.5 g/l	2140	480	367	5.5
	5,0 g/l	3450	590	420	5.7
	10,0 g/l	2930	770	796	6.7

because for lowering the level of toxic ions or pesticides to normal concentrations the required size of a plateau will be too big.

Of all the methods studied of drainage water treatment some are too expensive, others are insufficiently researched and tested under full-scale conditions. Desalinization technology is applied in many countries for after- treatment of natural water in water mains, treatment of wastewater and desalinization of saline and marine water. As for desalinization of drainage water, only in the U.S. has there been experience of that. Drainage water from irrigated lands in the Mojave desert go down along the Gile river to the River Colorado polluting its water which stirred protests of Mexican states. For correction of the situation and improvement of the quality of river water in Yuma (Arizona), they constructed a desalinization plant for treatment of drainage water. About 124 million m³ of desalinized water (0.2 g/l) is diverted into the Colorado river, and after mixing with river water they are used in Mexico for irrigation.

18.8 The Desert Institute's New Technology

The Desert Institute developed a technology for the removal of pesticides from drainage water by applying natural sorbents in the form of swelled argillites. Adopting this technology "*Turkmengiprovodkhoz*" prepared several projects of sorbtion plants for treatment of drainage water in Turkmenistan: in Nebitdag with a production capacity of 0.3 m^3/s; in the Ashgabat etrape – 0.6 m^3/s; Takhta-Bazar etrape – 0.6 m^3/s and others. After treatment the drainage water should be used for irrigation. According to the authors' estimates, the cost of drainage water treatment on these plants should be less than other treatment methods.

One should distinguish traditional desalinization (option II), in which from saline water they recovered 50-70% of fresh water and 30-50% of effluents but then all salts of the original water are disposed of into the natural environment damaging it. Another approach, that excludes the possibility of environment pollution and is consistent with modern ecological requirements, should be taken in treatment of drainage water. That is why the Desert Institute developed and proposed a new, more progressive technology: integrated demineralization of drainage water (option III) which incorporates water conditioning, desalinization, concentration of wastewater, and the processing of concentrated salt liquors with the recovery of commercial salts and dry residue (Jarkov 1995).

After study and selection of available technological equipment for various processes of integrated demineralization of drainage water, they developed schemes for three desalinization technologies (distillation, electrodialysis and reverse osmosis).

Using the existing prices of the selected equipment and the enlarged costs of construction of buildings, reservoirs and communication methods, capital costs were estimated (without costs of land and taxes on capital). Applying world prices on energy, reagents and materials the operational costs taking into account maintenance costs were calculated. For specific consumption of energy in desalinization and salt liquor treatment, the total energy use was estimated expressed as conventional fuel (calorific value 7,000 Kcal/kg or 29.3 MJ/kg), including the consumed electric power. Table 18.2. gives the estimated economic indices and energy use for traditional desalinization and complex demineralization when additional fuel is used in the concentration of wastewater, processing of salt liquors, and evaporation of tails to a dry state. The same Table 18.2. gives technical-economic indices for options II and III of water conservation engineering routes. In order to get the necessary data for a particular case, one needs to multiply figures from Table 18.2. by the volume of treated drainage water. The table shows that it is advisable to treat low saline drainage water applying membrane methods, and highly-saline drainage water using distillation methods.

The Desert Institute investigated and assessed possibilities for desalinization of a drainage flow for conservation purposes. Desalinization is called to resolve one more problem: that of obtaining additional fresh water resources. As a result of reconstruction of irrigation systems, it is planned by 2005 to have diminished the drainage flow to 4.15 km^3, of which the most heavily polluted part will be 1.85 km^3. Treatment of such amount of drainage water will require construction of an integrated, demineralization plant with a capacity of 5,236,000 m^3/day (US$ 5.14 billion), and to burn annually 23 million tons of conventional fuel. This will provide 1.7 km^3 of fresh water and 16.2 million tons of different chemical products.

The above-mentioned plant in Yuma (USA) is designed to desalinate drainage water only for environmental conservation purposes. The Desert Institute suggested quite a new idea of drainage water desalinization for purposes of water supply, irrigation and environment protection from pollution. It is proposed to construct an agrodesalinization plant in the Boldumsazsky etrape of the Tashauz velajate. It will process nearly 600,000 m^3/year of drainage water from the Geok-Chaga collector. This plant will undertake integrated demineralization of drainage water, the high-quality desalinated water will be used for drinking water supply, water of worse quality will be used for irrigation and growing of ecologically pure products.

If the problem of the use of desalinated water for water supply has been adequately studied, then the problem of its use for irrigation has not been studied under full-scale conditions. Laboratory studies have shown that in the course of desalinization of drainage water when applying a membrane technology (reverse osmosis and electrodialysis), water becomes enriched in single-valence ions of sodium and chlorine. That is why when using desalinated water for drinking water supply or for irrigation, it should be subject to conditioning.

18.9 Estimates for Proposed Desalination Complex

The proposed agrodesalination complex should provide a big village or semi-urban settlement (50 l/day per a person) with water whose quality is consistent with international standards, and with water for growing ecologically safe products for baby feed. Rough estimations of such a project will include:

1. Desalinization complex

1.1. Unit for drainage water conditioning (salinity 5 g/l) with a capacity of 1,600 m^3/day, specific capital costs \$ 200 per m^3/day – total cost \$ 320,000.

1.2. Desalinization plant with a capacity of 1,000 m^3/day, specific capital costs \$ 1,000 per m^3/day – total cost \$ 1 million.

1.3. Station for drinking water with a capacity of 1,000 m^3/day, specific capital costs \$ 200 per m^3/day – total cost \$ 0.2 million.

1.4. Unit for concentration of effluents from a desalinization plant with a capacity of 600 m^3/day, specific capital costs \$ 1,200 per m^3/day – total cost \$ 0.72 million. This unit will provide 550 m^3/day of concentrated salt liquor with a salinity of 100-120 g/l.

1.5. Unit for salt liquor processing with a capacity of 50 m^3/day, specific capital costs $ 1,600 per m^3/day – total cost $ 0.08 million.

2. Agroindustrial complex

2.1. Greenhouses with an area of 5 ha, specific capital costs $ 200,000/ha — total cost $ 1 million.

2.2. Open ground covering an area of 30 ha equipped with up to date irrigation facilities (drip and subsurface irrigation), specific costs $ 10,000/ha – total cost $ 0.3 million.

2.3. Administrative buildings, laboratory and dwellings, plus warehouses with a total area of 2,000, specific costs $ 50/m^2 – total cost $ 100,000.

2.4. Auxiliary structures (reservoirs, communication, distribution network, devices, reinforcements, etc.) – total cost $ 0.54 million.

The total approximate cost of the agrodesalination complex will be US$ 4.3-4.5 million. For this the pilot project will be a demonstration plant giving an example of a wise attitude to the environment, rational management and non-depleting use of land and water resources applying new, farming practices.

References

Glazovski NF (1990) Aral disaster (in Russian). Nauka, Moscow.

Razakov RM (1991) Utilization and protection of water resources in Central Asia (in Russian). Mekhnat, Tashkent.

Sanin MK *et. al.* (1991) Sarykamysh lake and reservoirs-accumulators of collector – drainage water (in Russian). Nauka, Moscow.

19. The Drinking Water Supply

V.V. Jarkov

ABSTRACT: Water in the Central Asian republics is scarce, and much of the old hydro infrastructure is antiquated and unsatisfactory. The Desert Institute has addressed the pressing problem of water treatment and developed inexpensive and effective, new devices. Of central importance in this work has been the testing of new sorbents to absorb the water pollutants. These portable units have been able to meet the enumerated goals set both for ordinary households and for industrial purposes.

Countries of Central Asia suffer from an acute deficit of fresh water supplies. The existing water sources (rivers, canals, collectors, and groundwaters) are intensively polluted with chemicals (pesticides, herbicides, defoliants, and heavy metals). Apart from direct deterioration of the water quality, the sources of water supply are polluted with wastewaters and drainage waters as well as other wastes of agricultural and industrial production. Nevertheless, insufficient attention as yet has been drawn to the study of water quality in the region. Deteriorating water quality, growing water salinity, worsening water use conditions and poor sanitary state of settlements are the main reasons for deteriorating conditions of life and health of the population.

A reliable way to remove hydrophobic pesticides from water is the intensification of coagulation processes which requires great quantities of activated carbon. As there was no activated carbon, a search for new kinds of sorbtion materials and their employment in practice made it possible to introduce a sorbtion method to clear water mains, and widen the production and application of portable, water treatment devices.

19.1 Why Existing Facilities Fail

In Turkmenistan there are 295 water mains, of which 107 are fed from surface water sources and 188 from groundwater. Of all the water mains 49.8% do not comply with the sanitary-technical requirements; 20.3% have no treatment facilities (completely or partially); 22.5% do not have water disinfection units; 25% of water main systems have an expired depreciation period, and many accidents happen here.

Traditional treatment facilities on water supply stations are not a reliable barrier on the way of pesticides to treated water. Water treatment facilities do not provide protection from many pesticides: dinitrophenol and ditrophoorganic compounds, derivates of carbamic, thio and dithiocarbamic acides, derivatives of urine and others. Satisfactory results from water treatment following a traditional scheme can be awaited only for substances not readily soluble present in the form of suspensions or particles in a colloidal degree of dispersion.

Such an unsatisfactory state of water supply facilities is responsible for the low quality of drinking water. A specific share of non-standard samples for sanitary-bacteriological parameters (coli-index) makes up annually 28.2% and for sanitary-chemical parameters 25.3%, i.e. epidemiologically dangerous water is supplied to a considerable part of the population. Water quality at water intakes is very poor. In the period of the highest water consumption the coli-index reaches 240,000, consequently, such water requires special treatment and disinfection.

The problem of delivering drinking water of good quality to the population is acute both for small communities taking water from unequipped water sources, and for the wider population provided with centralized water supply, because a drastic deterioration of the ecological situation often leads to a situation when water in surface water bodies by its chemical composition and microbiological parameters is often close to wastewaters.

Keeping this in mind the traditional systems for water treatment applied on water supply stations cannot treat polluted water to a drinking standard, because they cannot remove many chemical substances. At the same time in the course of water treatment and disinfection, the level of halogen-containing compounds undergoes a multiple increase. These compounds are formed from organic pollutants after they enter into a reaction with chlorine, they produce mutagenic and carcinogenetic action and affect the water user. So an admissible concentration of these compounds in drinking water of some water supply stations exceeds MAC (maximum allowable concentration) by 100 times.

19.2 The Necessity of Portable Units

Modification of the existing public facilities and installation of new ones for after-treatment of drinking water to bring its quality to a required level are very costly, and they cannot often be provided within a short time period. Besides, there are many users who draw water from private wells or via pipelines that do not provide adequate water treatment.

In our opinion an optimal method of water treatment is a wide distribution of portable units that are designed:

- to disinfect water and remove microorganisms to a level safe for a man;
- to purify it from mechanical admixtures and soluble toxic substances;
- to remove alien taste and smell;
- to ensure supply of drinking water of good quality;
- to be completely autonomous and to comply with ergonomic possibilities;
- to have small dimensions and weight that will make it applicable in any conditions;
- to be constantly ready for use;
- to ensure supply of treated and disinfected water within a short time period.

19.3 New Sorbents from the Desert Institute

For these purposes the Desert Institute developed new kinds of sorbents: modified argillite, sulfocarbon, bentonite, bactericidal sorbents and their modifications that were tested in Turkmenistan (Jarkov 1995). In Ashgabad water from a water main was used in investigations. Water was fed via a special unit to treatment units filled with modified argillite, bentonite sorbents, and sulfocarbon. These were compared with BAU-carbon (birch-active carbon). After each 500 liters of water that passed via sorbents, they took samples and analyzed them.

19.4 Test Results

Applying an extraction photometric method, they determined PAV and IR, while with the spectrographic method - petroleum products, with boronmetric and spectrophotometric methods (η-nitroaniline) - single-atom volatile phenols.

Investigations have indicated that PAV in water are revealed very rarely, and from the obtained results it is impossible to make certain conclusions. BAU-coal is good at removal of petroleum products with a retention percentage if 65-70%, application of modified argillite and sulfocarbon gives 50 to 100%. Bentonite sorbent works badly with petroleum products.

The above sorbents do not affect single-atom, volatile phenols with their low content in water.

Investigations also showed that BAU-carbon and argillite as well as a three-layer complex of sorbents taken in equal quantities (modified argillite, sulfocarbon and BAU-carbon) are good at absorbing copper. A bentonite sorbent absorbs copper to a lesser degree, although the retention percentage is 20-55%.

The best sorbing capacity in respect to manganese was revealed by a bentonite sorbent, the retention percentage of which is 55-60%. A complex of three sorbents absorbes copper well, its retention percentage is 60-96%. Argillite modified sorbent performs poorly with manganese.

A complex of three sorbents shows good absorption of zinc over a long time period; the retention percentage is 45-60%. Modified argillite is good in absorbing zinc from the first 1,000 liters of water, later on its sorption capacity drastically drops.

Modified argillite sorbs iron well from the first 2,000 liters of water, but as the Ashgabad water has a high iron content, the sorbent will be soon saturated. The remaining sorbents displayed a weaker capacity, their retention percentage was only 20%.

Water samples were subjected to analysis for a level of rated micro- and macrocomponents (cobalt, lead, cadmium, and nickel) applying a method of atomic adsorption. Investigations have indicated that BAU-coal absorbs well, while all the other sorbents are satisfactory; their retention percentage is 20-30%.

A complex of three sorbents achieves a retention percentage of 85%, and BAU-coal, bentonite and argillite are able to achieve up to 75% in sorbing lead.

Analyses have shown that all the above sorbents absorb cadmium well at the initial stage, and the retention percentage is 75%.

It is seen from the obtained results that coal, more specifically BAU-coal as well as modified sulfocoal entering a complex are good at removing chromiun from water.

The content of lithium in tap water is constant, and is equal to 0.1033-0.1123 mg/l. Sorbents do not influence its content except for a three-layered screen from modified argillite, sulfocoal and BAU-coal. Most likely sulfocoal is critical here.

The level of calcium in water is steady, varying from 64.17 to 86.15 mg/l. The studied sorbents do not influence it perceptibly, and the same is true of magnesium ions (from 50.22 to 63.84 mg/l).

Investigations have shown that sorbents do not influence the pH of water.

Oxidation (permanganate and bichromatic) is an important index establishing the presence of pesticides in water. It is expressed in the amount (mg/l) of oxygen that is necessary for oxidation of organic substances (pesticides) in water. Permanganate oxidation reveals the presence in water of readily oxidizing substances, while bichromatic oxidation reveals poorly oxidizing substances. Experiments have revealed that heavy organic substances are well trapped by BAU-coal, bentonite, the retention percentage is equal to 55-60%, while light substances are trapped modified sulfocoal (68-70%).

Analysis of water samples for pesticides applying a chromatographic method have shown that BAU-coal, modified sulfocoal and a three-layered component absorb practically completely α GHCG (hexa chlor cyclo hexan) (98-100%). The above sorbents absorb β GHCG well too. But modified argillite removes γ GHCG from water between 25-50%.

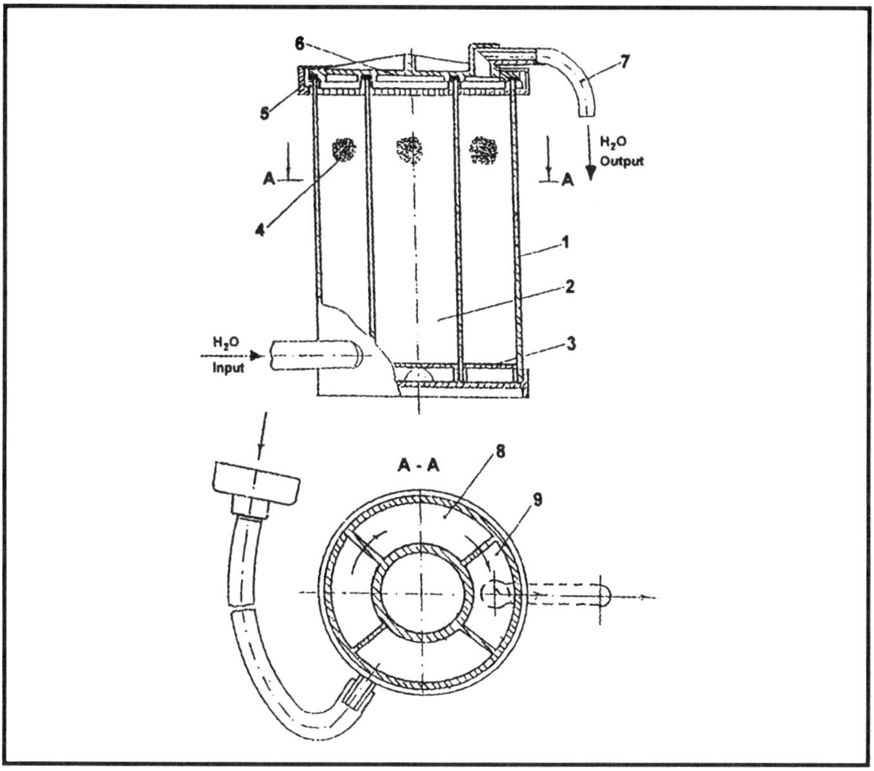

Figure 19.1. "Cheshme-3" domestic device for drinking water refinement **(1)** frame **(2)** central section **(3)** netting **(4)** sorbent padding **(5)** - screw **(6)** cover **(7)** flowing hose **(8)** large section **(9)** small section

19.5 Water Treatments and Filter Systems Developed by the Desert Institute

19.5.1 The Water Treatment Unit "Cheshme"

On the basis of the above sorbents they developed a water treatment unit "Cheshme" (Jarkov *et. al.* 1992) which purifies dringking water of chemical pollutants (pesticides, heavy metals, phenols, radionuclides, etc.). The water treatment unit “Cheshme” (Fig. 19.1.) uses such sorbing materials as BAU-coal (GOST 6217-74), modified argillite sorbent consisting of gravel and thermoargillite sand (swelled argillite, RST (standard) of Turkmenistan 427-85 obtained after thermal treatment of crushed argillite of the Greater Balkhan in Turkmenistan; a modified argillite sorbent is patented, Author's Certificate No. 1603709, class CO2 1/28 and specification TU-303-90 UVEI 060.711.001 TU agreed upon with the Director of the A.N. Sysin Research Institute of General and Communal Hygiene 20.02.1991), modified sulfocoal. In respect to this sorbent an application for the proposed invention has been submitted No. 49337656/26a31470 and a specification agreed upon (TU-303-90 UVEI 060.711.002 TU). The former USSR Ministry of Health issued permission for use of the modified sorbents argillite and sulfocoal as a filtration charge during treatment of drinking water from chemical pollutants.

The water treatment unit “Cheshme” weighs 3.5-4.0 kg. The maximum admissible water pressure is 1000 hPa (1 kg/cm^2). The figures for removal of pesticides are as follows: (DDT [dichlordiphenyltrichlormethylethane], for DDD [dichlordiphenyldichlormethylmethane], for DDE [dichlordiphenyldichlorethylene], GHCG, metaphos, sevin and others) - 80-96%, heavy metals (copper, lead, boron, cadmium, zinc and others) - 90-99%, phenols - 96-99.9%, iron - 50-60%, and for manganese - 80-90%. An optimal production capacity is 5 l/day. The working lifetime of sorbents at optimal production capacity is 18,000 liters.

The “Cheshme” unit is less costly than water treatment units manufactured outside CIS states.

19.5.2 An Affordable Filter System FOPV-1

A filter for treatment of drinking water (FOPV-1) is intended for after-treatment of water through removal of chemical pollutants. It can be used in domestic, drinking water supply systems, in pressurized water supply networks for individual and group use for drinking purposes (Fig. 19.2.).

This filter is best suited for regions with a poor ecological state of water resources. It removes pesticides, toxins, heavy metals, radionuclides, and partially nitrates and nitrites from water. The weight of FOPV-1 is 200 g.

This filter is connected to a tap with a cone end and a collar. It is kept on a tap due to friction and does not require additional fixing. The maximum pressure in a pipeline is 2,000 to 5,000 hPa.

Which filtering element to use and their lifetime depend on the degree of water pollution.

The filters use as filtering elements the pressed sorbents in the form of tablets permitted for use by the USSR Health Ministry.

Sanitary and hygiene tests of FOPV-1 were conducted at the laboratory of the drinking water supply of the A.N. Sysin Institute. They tested a chemical composition of a filtering element; studied the possible migration of chemical substances

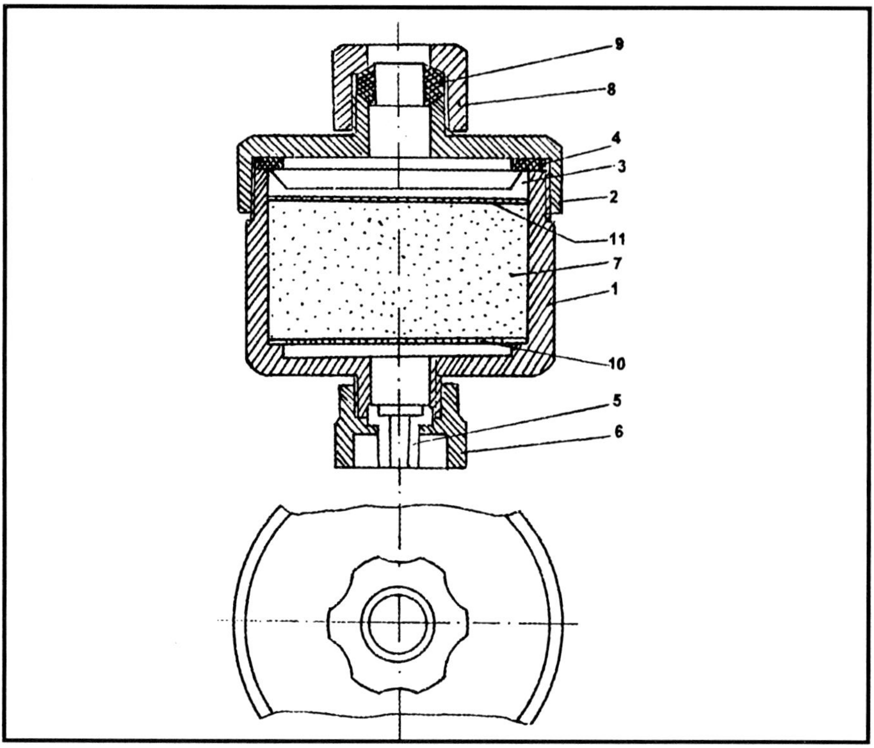

Figure 19.2. Filter of FOPV-1 water refinement **(1)** frame **(2)** cover **(3)** ring **(4)** rubber padding **(5)** carbine **(6, 8)** screw **(7)** filtering element **(9)** stuffing box **(10, 11)** netting

from a filter under aggravated conditions in a through-flow mode; studied the efficiency of after-treatment of water and a capacity to remove heavy and alkali-earth metals, halogen-containing elements, chlor- and phosphororganic compounds, phenols, PAV (active substances), nitrates, nitrites, integral indices of organic pollution; multipurpose appraisal of water quality additionally treated by the filter using data of biotests on water test-objects; studied a water quality after additional treatment by FOPV-1 by microbiological indices in service-life tests.

A special analysis of two types of a filtering element made on the basis of the sorbents BAU + modified argillite and BAU + bactericidal bentonite has indicated that unlike BAU the sorbent on the basis of BAU + modified argillite retainsa much higher content of silicon (10%), aluminium (10%), iron (10%), manganese ($6x10^{-2}$ %), copper ($8x10^{-3}$ %). A bactericidal bentonite sorbent is added to the sorbent tablet so that no bacteria develop in BAU-coal.

A study of a content of polycyclic, aromatic hydrocarbons (PAH) in water extracts has shown that there is no migration of carcenogenic substances (benzpyrenes) from the studied sorbents.

Studies of the water treatment efficiency in service-life tests (1000 l) by removal of different pollutants (heavy and alkali-earth metals, halophor compounds, pesticides, nitrogen-containing compounds, and phenols) were carried in every 100 liters of water at a filtration rate of 200 ml/min.

It was found that for water pollution with chlororganic compounds (GHCH, DDT at a level of 1-2 MAC) the percentage of their retention for studied filters was on the average not more than 28-43% (GHCH) and 24-58% (DDT), although a retention percentage of 66-75.3% was also observed. The efficiency of water purification from phosphororganic pesticides was somewhat lower and varied from 28.6-30.6 and 26-30%.

A high efficiency of purification from aluminium and iron, irrespective of a composition (100-1000 l) was found at filtration up to 500 liters of water (75-95%), then it dropped and after 1000 liters was equal to 25%.

19.5.3
The Filtering Unit "Zha-Ro"

We have also developed a filtering unit "Zha-Ro" intended for additional purification of drinking water from microbiological and chemical pollutants; it can be used in domestic, drinking water supply. The body of the "Zha-Ro" unit consists of two cylinders inserted in each other, while between them a special bactericidal, bentonite sorbent is placed.

The unit is closed with a sealing cover from which runs a nozzle that is pulled on a tap. A cover, ring and a gasket between it and cylinders are manufactured from materials permitted for use in domestic and drinking water supply. Those cylinders are made of polymer concrete and Karakum sand is used as a filler here. The weight of the bentonite sorbent is 60 gramm. The sorbent has the form of pressed pellets "worms" 11 mm in diameter and 5-10 mm long.

Efficiency of water treatment with a bactericidal, bentonite sorbent was assessed at various contact times. Tests were conducted at the Laboratory of Virusology of the A.N. Sysin Institute of Hygiene. The obtained results showed good efficiency of disinfection by the bactericidal, bentonite sorbent by a total microbe number. The efficiency reached 99.7% for a contact time 10 minutes and more. The coli-index went down to zero during half an hour contact.

Application of a small quantity of the bactericidal bentonite sorbent in "Zha-Ro" have revealed its fairly low sorption capacity. But still it retained pesticides well. The retention percentage in respect to GHCH was 34.

The bentonite sorbent used in the "Zha-Ro" unit has displayed a high sorbing capacity in respect to large and mostly hydrated ions of metals. A retention efficiency of aluminium, iron and manganese was 50-80%. Absorption was ensured not by the sorption volume of bentonite, but rather by the finely-porous body of the "Zha-Ro" unit made of small-grained Karakum sand. The retaining capacity of a body of the unit in respect to some metals is good: copper by 35%, molibdenum by 40%, zinc by 16%.

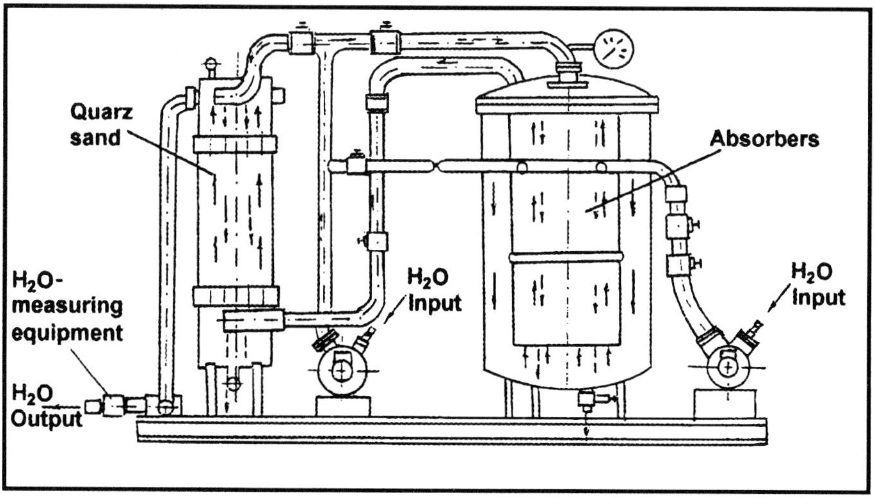

Figure 19.3. Technical scheme of water treatment unit VOU-2.5

19.5.4
The Water Treatment Unit VOU-2.5

The Desert Institute developed a water treatment unit VOU-2.5 (Fig. 19.3.) which is manufactured at the Design Technical Bureau of "Agroprom". Its capacity is 2.5 m³/h. This unit is intended for purification of water from chemical pollutants.

The unit VOU-2.5 consists of:

- a filter which is a vertical overturned cylinder 700 mm in diameter divided into two sections, the upper is filled with BAU-coal and the lower with modified sulfocarbon, the volume of each section 0.2 m³;
- the upper distribution unit intended for feeding water into a filter and which is represented by a collector with outlets directed upwards;
- a clarification chamber consisting of cartridges filled with quartz sand and the upper and lower distribution chambers;
- the lower distribution unit intended for water diversion during filter flushing;
- two centrifugal pumps for feeding of initial and flushing water.

The unit is operated as follows:

- Water from a water source (basin, well, water tower) is fed by a pump or by gravity under a pressure of 5,000-10,000 hPa via the upper distribution unit to a filter, is spread over a surface, passes through a sorbent layer upwards, and is delivered along a pipeline to a chamber for clarification; and then to a user.

- With the exhaustion and contamination of a filter and a clarification chamber, and after exceeding of the maximum, admissible levels of harmful, chemical elements in the filtered water the filter is subject to back-flushing.

  If a putrid smell of water appears the sorbent should be replaced.

The water treatment unit VOU-2.5 was tested in its first year in the settlement Enev, in the second year in premises belonging to the Turkmen Desert Research Institute base. Water was taken from the Karakum canal. In the first year of operation VOU-2.5 ensured the complete removal of copper, zinc and iron from water. Investigations have shown that water containing dust at a rate of 0.184-0.233 μkg/l was treated by the unit completely (only traces of this substance were found in a filtrate). A content of α GHCG was reduced by 4,500%, and of γ GHCG 1,500%.

Ordinary water (a counter reading 570 m^3) was polluted with more widespread pesticides: nitrophenol at the rate of 0.045 μkg/l, and magnesium chlorite at 0.0456 μkg/l. After chromatographic tests of water at the laboratory of the Desert Institute, these pesticides were not found. Investigations have indicated that water for some indices is treated well, for instance, a solid residue decreased from 567 μkg/l to 475 μkg/l.

The water treatment unit VOU-2.5 was tested for purification from bacteria according to GOST 18963-73. The results have shown that the content of bacteria in water before treatment and after it remained the same.

The capability of the sorbent to retain chemical elements in relation to an inside pressure was examined. It was found that a treatment efficiency of the unit is the highest at a pressure of 100-500 hPa.

The main technical characteristics of the water treatment unit VOU-2.5:

- rated production capacity: 2.5 m^3/h
- minimum production capacity: 1.0 m^3/h
- internal diameter of a filter: 700 mm
- operational pressure: not more then 1,000 hPa
- filtration area: 0.19 m^2
- sorbent volume: 0.4 m^3
- dimensions: not more than 1930 * 1045 * 2100 mm
- weight: 1200 kg.

Apart from a sorption capacity of filters, it is necessary to know the threshold values of the different substances in water to be filtered so that the unit can remove them sufficiently to comply with standards on drinking water. In experiments original water was intentionally polluted with heavy metals (Pb, Cu, Ba) and pesticides (DDT, DDE, DDD, α GHCG, γ GHCG) in concentrations 5; 7.5; 15 MAC. Contaminated water was circulated through the unit under different operational modes. It was found that the optimal capacity of the unit proceeding from a purification capacity is 1 cu. m/h. It is at this capacity that the required contact time (9.5-11.3 min) of treated water with sorbent is ensured at a filtration rate 5.3-6.3 m/h.

Therefore, for the tested filter the maximum concentration of pollutants in original water at a capacity of 1 m^3/h can reach fifteen-fold and more values of maximum admissible concentrations.

This filter can be used only for purification of water from organic substances. Its place in a technological scheme should be after the clarification (sandy or high-rate) filter. Sorbents are used only for design purposes i.e. for removal of chemical pollutants. The sorbent receives clarified water that is why it needs flushing seldom without unnecessary losses of coal. Clogging of coal with suspended matter is insignificant which creates good conditions for sorption of chemical pollutants, and prolongs the service life of coal as a sorbent.

Consequently, in order to observe the conditions for normal operation of a filter with a coal charge, it is necessary that water that goes to it should meet the requirements of GOST 2874-82 "Water drinking" for its suspension content i.e. its turbidity should be 1.5 mg/l.

19.5.5 Industrial Filtering Plants

Results obtained in the course of experiments with the VOU-2.5 unit made it possible to design industrial plants. Having the approval of the Turkmenistan Health Ministry for application of the developed sorbents for treatment of drinking water, and applying materials permitted for use in the drinking water supply, we developed plants for additional treatment of drinking water UOV-3, UOV-5 (Fig. 19.4.), and UOV-10. These plants are intended for mechanical and physico-mechanical, additional purification of drinking water from different pollutants. These can be finely-dispersed, suspended organic and mineral substances, pesticides, salts of heavy metals, partially nitrogen and phosphorus-containing compounds, phenols, radionuclides and others.

The plants should be connected to a water supply network with a pressure from 100-800 hPa, and a maximum flow rate 3.5 to 10 m^3/h, respectively. They are designed to supply drinking water to separate regions, settlements, high buildings, enterprises, schools and hospitals.

These plants consists of several tank filters in which original water is directed from below through sprayers which ensures dispersed passage of water through a sorbent and prevents its washing out. Each filtering tank is filled with one of the sorbents.

After contamination and partial depletion of a filtering charge, flushing is carried out. Regeneration and clearing of sorbents are made separately in each filter. For this purpose the upper diversion tube is disconnected, and instead of it a flexible hose (arm) is set. Valves on inlet and outlet from a filter should be closed, and a diversion coupling should be connected to a sewerage pipe. Flushing water going under pressure through a sorbent removes pollutants. Then a flexible hose is connected to a second filter, valves should be also closed, and the operation is

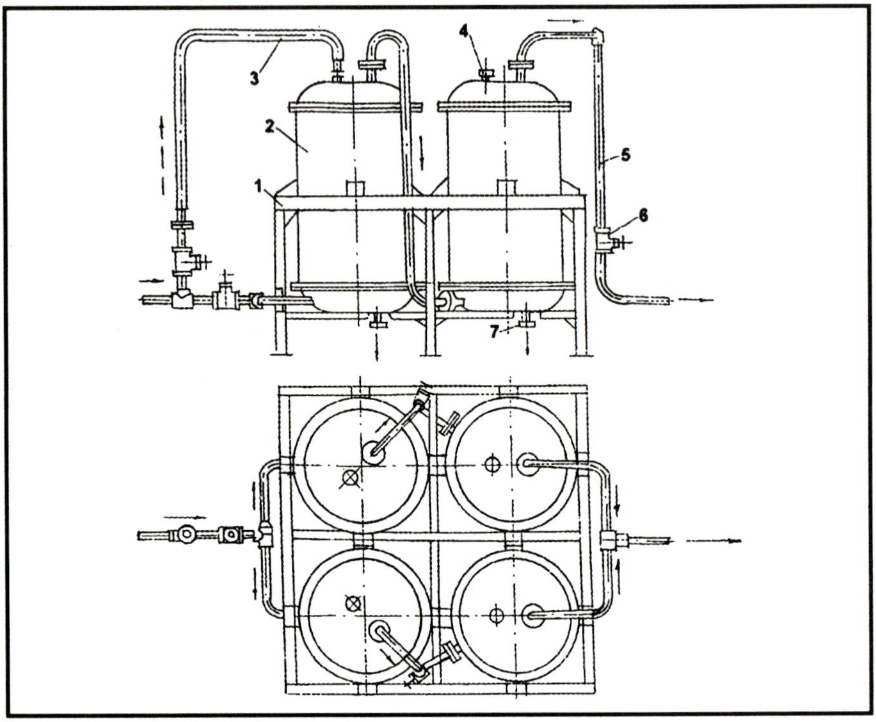

Figure 19.4. Technical scheme of water refinement device UOV-5 **(1)** frame **(2)** filter **(3)** flexible hose **(4)** air-valve **(5)** pipelines **(6)** valve **(7)** carbine

repeated. The same with the third and other filters. Flushing should be performed at a rate several times faster than the operational rate during filtering.

One of the tanks is filled with BAU coal. Coal regeneration is not always economically justified, especially in view of the fact that every year fresh coal should be added in the unit to compensate for its losses for crushing, wear out during flushings. At the same time as a result of overgrowing of coal with inorganic pollutants (mostly with aluminum, iron hydroxides and others) its sorption capacity in respect of organic substances can sharply deteriorate. Therefore, it is necessary to ensure very efficient clarification of water before it enters into coal charge.

19.6 Conclusion

Developed technologies for getting new kinds of sorbents from local materials in Turkmenistan, individual water treatment units and industrial plants will make it possible to provide Central Asia with good drinking water.

References

Jarkov VV (1995) Improvment of drinking water quality. Probl. Osv. Pustyn 1995, No. 4: p. 69-81.

Jarkov VV, Jarkov DV, Filippova AV (1992) Domestic apparatus "Chesme" for drinking water cleaning. Ashkabad.

20. The Application of Solar and Wind Energy

M.V. Kolodin, S.S. Seitkurbanov, K.C. Charyev

ABSTRACT: Sustainable energy resources have long been of experimental significance for the Desert Institute especially solar and wind power. Different types of plants have been developed to harness these natural energy sources including a combined helio-wind plant. The experiences so far are described.. The use of osmotic membranes has been especially successful. Of particular significance for this work was their use to desalinize water for irrigation.

The most stable, energy source on the Earth is solar radiation. At the Ashgabat latitude the sum of solar radiation (direct and diffused) on a perpendicular in respect to a ray's surface is 6930 MJ/m² a year. If an average figure for Turkmenistan is 6250 MJ/m² a year, then the whole territory of the country gets annually 850 x 10^{12} kWh of solar radiation.

Energy of air currents through the whole atmosphere is 2-3% of the solar heat flow and is equal nearly to 20 x 10^{12} kWh. But a man with the help of windmills can use wind energy only within a thin terrestrial layer (10-100 m), and potential wind-energy resources on the territory of Turkmenistan are estimated at 640 x 10^9 kWh, which is 1330 times less than a solar heat flow (Kolodin 1992).

Solar energy gets to the Turkmenistan territory fairly evenly, slightly increasing in the Central Karakum and in the south-east of the country (7,000-7,400 MJ/m² a year). Practically everywhere in Turkmenistan there are favorable conditions for harnessing solar energy.

Wind energy is distinguished by great unevenness, and so in the 1970s the Desert Institute investigated wind-energy thoroughly, classified recurrence of wind speeds, estimated the likely power generation and expected idling period of windmills, and presented these data in tables and on maps (Kolodin 1992). A map of wind energy resources characterizes gross wind energy during a year through 1 sq. m of a surface recorded by a windpropeller. In order to estimate useful energy, it is necessary to multiply gross energy with the coefficient of wind energy use and the efficiency of a machine (electric motor, pump, mill, etc.).

It was found that wind intensity is lowest in southern piedmont regions where gross energy was 100-200 kWh, in the south-east 200-400 kWh, in the Central Karakum 400-600 kWh, in the western regions 600-800 kWh and in the Caspian region 800-1000 kWh. These data allow calculations of the likely energy generation by a windmill in any region of the country.

20.1 Energy Needs for More Water Desalination

In Turkmenistan all river waters are used mostly for irrigation in a cultural zone, while the desert territory (more than 80% of the area) is devoid of sources of surface waters, and the only source of water there are groundwaters mined by wells and pits, and water from atmospheric precipitation. More than half of all water sources feature a water salinity over 10 mg/l, which makes this water unsuitable for cattle drinking. For the development and water supply of waterless pastures, cattle breeders have for long collected water on takyrs. More recently, water has been brought here in tank trucks, but this is very expensive especially if water has to be transported to distant pastures.

So, other sources for the water supply of pastures should be sought. In order to get fresh water on site, groundwaters should be desalinated, and this requires much energy. Pastures do not have any energy base. Scientists and engineers started the development of recommendations on the use of non-traditional and renewable energy sources for water desalinization.

20.2 Earliest Attempts to Harness Solar Energy

The first investigations were undertaken by the Scientific-Production Association "Solntse" ("Sun") of the Turkmen Academy of Sciences on desalinization of saline waters in the simplest, solar desalinization units of a greenhouse type (Bairamov and Seitkurbanov 1977). In the Bakhardensky etrape of the Akhalsky velajate in the late 1960s and early '70s, two desalinization plants were built on desert pastures in the small settlement Oves-Shikh with a capacity of 4 m^3/day (an area of 800 m^2), and Cherkezli with a capacity of 6 m^3/day (an area of 1220 m^2) which provided water for 800-1000 sheep.

Desalinization plants were assembled from prefabricated, glazed and reinforced concrete sections with an area of 19 m^2. Their design envisaged a possibility of collecting atmospheric precipitation from a glazed surface. The maximum specific capacity of 1 m^2 of this plant was 5 l/m^2 a day. Its cost was US $ 56-65. The cost of 1 m^3 of the obtained distillate was $ 4.2-5.0. A low capacity and high material consumption (more than 300 kg/m^2) did not add to the popularity of these desalting plants. Apart from this, its operation has revealed that reinforced flumes withstand being used in conditions of an aggressive medium (salt liquor concentration of 120-250 g/l) for more than 14 years; after this period they are not usable.

So, other, cheaper and more corrosion-resistant plastic materials (polypropylene, faolite and others) are necessary. Researches along this line are continuing. In other countries an interest in desalting plants of the greenhouse type declined, and solar energy is used in plants with other improved, desalinization processes (Kolodin 1985).

The Turkmen Branch of the Research-Production Association "Kvant" (Moscow) developed several types of solar photoelectric stations (SFES) for agricultural use: for electrification of the houses of shepherds, for lifting and desalinization of water on pastures. The first solar station SFES-220 with a capacity 200 W was built for mechanized lift of water from well Iner-Chaga in the Ashgabad etrape. A solar water lift proved more efficient than a strip water lift VLM-100 with a petrol engine ZID-4.5. The cost of water lift with the help of solar energy is US$ 0.29/m^3 instead of US$ 2.55/m^3.

A more powerful, photoelelctric station FV-500 with a capacity of 500 W equipped with a device for automatically following the sun was constructed at the same farm on the well Teze-Gui. Of the total energy generated by this plant, 200 W were used for driving a pump "Malysh", and 280 W for energy supply of the electrodialysis, desalinization plant with an output of 0.6 m^3/day. The cost of desalinated water was US$ 7.15/m^3. The pump daily lifted about 4 m^3 of water, and the cost of 1 cu.m of lifted water was US$ 0.57, that is cheaper in comparison to a petrol engine (US$ 2.55/m^3). An annual economic effect of application of a solar plant was US$ 2,720 and construction costs were repaid during 6 years.

20.3 The Institute's Work on Wind Energy

The Desert Institute carried out researches into the application of wind energy for desalinization of water to ensure water supply of pastures. On many pastures cattle are provided with drinking water from wells where water salinity is 5-10 g/l, while water for shepherds and their families was brought from the cultural zone. For water supply to small users (shepherds, geologists, meteostations etc.), the Desert Institute together with the Moscow Research Institute of Plastic Materials developed and manufactured the electrodialysis desalinization plant "Barkhan" with an output of 30 l/h. It desalinated water with a salt content of 5 g/l and produced water of a drinking quality with a salinity 1 g/l. A bath of this electrodialysis unit is designed for successive distribution of water inside chambers. For this desalinization plant they produced the 12V electrodialysis unit, and it worked from a wind aggregate and a storage battery of 12V. Laboratory and field experiments have revealed certain engineering drawbacks.

Relying upon this experience a small portable wind-desalinization plant VOU-0.1 was developed. For this plant they designed a new, direct flow, electrodialyzer MPEO-1 with an output of 100 l/day. It incorporates frames with single-chamber, internal steps (a working part of a frame is divided into 4 sections); this made it possible to do without pumping, and it was enough only to lift a tank with original water with a capacity of 200 l to a height of 2 meters. This diminished energy consumption.

The power consumption of this desalinization plant was not high, and was equal to 15-20 W. VOU-0.1 derived its power from a wind-electrical unit VE-2M with a capacity of 100 W and during a calm period from a 12 V storage battery 6STE-180. The maximum salinity of treated water was 15 g/l, and that of the obtained dilution 1.5 g/l. At a lower salt content in original water, the quality of desalted water was better and power consumption decreased, too. Desalted water passed via a filter with activated coal into a tank 100 l in capacity. Tests of this desalting plant gave good results with desalinization of groundwaters and Caspian waters. The specific energy consumption was 3.0-3.3 KWh/m^3. During operation of the plant, salt depositions on membranes were detected, and so, special measures were developed to eliminate this. They tested intensive flushing of chambers, reversed polarization and acidification. For desert conditions, reversed polarization proved most advisable.

After the invention of osmotic membranes abroad, their application in water desalinization was very successful. When in the former USSR they started production of osmotic membranes, they were called hyperfiltration MGA-80 and MGA-90. It became possible to test them in desalinization of groundwaters. After laboratory and field experiments with these membranes, the Desert Institute developed on their basis two types of wind-desalinization hyperfiltration plants VOGU-1 for drinking water supply and VOGU-6 for pasture water supply.

In wind energy use, there can be two principal schemes:

- Scheme I: Accumulation of a ready product (fresh water) when a windmill and a desalinization plant are operated sporadically, and water is stored in a tank.
- Scheme II: Accumulation of energy when a windmill is operated sporadically and energy is stored in a storage battery, while a desalinization plant is operated to a certain schedule and provides a daily rate of water. The main technical and economic indices are given in Table 20.1.

It was expected that Scheme I would be preferable to Scheme II. However, estimations have revealed the advantages of Scheme II. The point is that a windmill is not operated all the time, and its energy-generating efficiency is described by a coefficient of installed capacity use which for the conditions of the Central Karakum is equal to 0.33. This means that the installed capacity of the windmill in Scheme I should be increased three times, and this increases capital costs and costs of desalted water (Table 20.1.). However, in desert conditions it is difficult to implement Scheme II (it requires a special room for storage batteries and more skilled service personnel). Therefore, Scheme I was applied.

Table 20.1. Technical and economic indices of wind desalinization plants for drinking (VOGU-1) and pasture (VOGU-6) water supply for conditions of the Central Karakum

Indices	Fresh water (I)		Energy (II)
	VOGU-1	VOGU-6	VOGU-6
Daily water consumption [m³]	1.0	6.0	6.0
Output of desalinization plant [m³/day]	3.0	18.0	6.0
Salinity of original water [g/l]	15.0	15.0	15.0
Salinity of desalted water [g/l]	1.0	5.0	5.0
Capacity of a fresh-water tank [m³]	6.0	36.0	6.0
Capacity of a storage battery [kWh]	2.58	2.58	45.78
Specific energy use [kWh/m³]	2.12	1.21	1.21
Power consumption [kW]	0.16	0.40	0.40
Wind plant capacity [kW]	0.48	1.20	1.20
Cost of a plant [US$]	4,980	20,780	12,290
Annual water output [m³]	329	1,971	1,971
Operational costs [US$]	1,160	6,010	5,450
Water cost [US$/ m³]	5.42	4.35	3.56

20.4 Combined Wind-Solar Energy

Considering the specific features of renewable energy sources the scientists try to develop methods of their use that will compensate the drawbacks of the other. In 1991 the Desert Institute suggested an idea for the combined use of solar and wind energy in a single isolated system (Seitkurbanov 1991). Solar and wind plants operated separately have for a long time been successfully working, but a system for combined use of wind and solar energy has not been conceived as yet by anybody.

As a result of theoretical investigations and the experience of previous researches, a major scheme was developed for a combined helio-wind energy generating plant (KGVES) for power supply to small users in a desert. At the Karrykul station of the Desert Institute in the Central Karakum an experimental plant was built that incorporates two, photovoltaic, solar power plants: FES-300 with a capacity of 300 W, and FES-400 with a capacity of 400 W; two windmill power generating plants VEU-0.12 with a capacity of 250 W; an inverter, a control block and an electrical measuring panel (Photo 20.1.).

Experiments have shown that the plant performed unfailingly. It was enough for one FES-300 and a windmill power generating plant. An inverter converted 12 V direct current into 220 V alternating current that provides lighting of a building, feeding of a radio set and TV. A 110/450 electric engine of a strip water lift was connected directly to the FES.

Photo 20.1. Combined helio-wind-power generation system

During October in tests the solar power plant reached a capacity of 561.3 W at a solar radiation of 783.3 W/m^2. The electric engine consumed 400-420 W of energy, while the capacity of a water lift was 700-720 l/h for lifting water from a well 17 m deep. Tests proved the engineering practicability of an idea to use solar and wind energy. Experience also showed that in order to ensure energy supply not only for domestic purposes and water lift, but for desalinization of drinking water for cattle the capacity of FES in a combined plant should be increased to 1.0-1.5 kW and a windmill, power generating plant to 3-4 kW. It is necessary also to take into account the accumulation of energy from renewable sources such as sun and wind.

References

Bairamov R, Seitkurbanov S (1977): Desalinisation of water by the hepl of solar power. Ylym, Ashgabat.

Kolodin MV (1992) Power resources of Karakum. Ashgabat. Ylym.1992: p. 157-171.

Kolodin MV (1985) Economy of water desalinisation. Nauka, Moscow.

Seitkurbanov S (1991) Combining helio-wind-power installations. Ylym, Ashgabat.

21. The Problems of the Aral and Caspian Sea

A.G. Babaev, C.O. Muradov

ABSTRACT: There are many pressing global environmental problems not the least serious of which are those of the Aral Sea and Caspian Sea (shrinking, pollution etc.). The Aral Sea has become increasingly salinized and desertified due to unwise exploitation and river pollution of two major tributaries, the Amudarya and the Syrdarya. The fluctuations in the Caspian Sea on the other hand have for a long time been scrutinized, and the various hypotheses to explain this phenomenon are listed and assessed. The authors argue that geological processes within the earth's crust are the likely cause. As with the Aral, further systematization of research on an international and regional level are required.

On the eve of the 21st century, the human species faces a number of vital challenges the most serious of which is aggravation of the ecological situation in arid regions. This is, first of all, due to the unprecedented development of desertification processes; as a result the situation is often a subject of discussions at different levels. This reflects a general anxiety of the world community about the future of the relations between nature and society. From year to year the number of ecologically imbalanced regions is growing, threatening the well-being of many millions of people. Among regions entering a zone of ecological disaster, a special place is taken by the Aral and Caspian Sea basins.

21.1 Aral Sea Problems

The Aral Sea basin covers the central part of the Turanian plain. The Aral catastrophe is a result of a predatory attitude to nature: when in search of the quickest benefits, decision-makers forget about the need to maintain ecological equilibrium.

In conditions of closed water basins, one of which is the Aral Sea, excessive water withdrawal leads to the desertification of river deltas and the sea water area.

The socio-economic development of the Aral region which was oriented to prioritizing exploitation of water-intensive branches of the economy had impacted

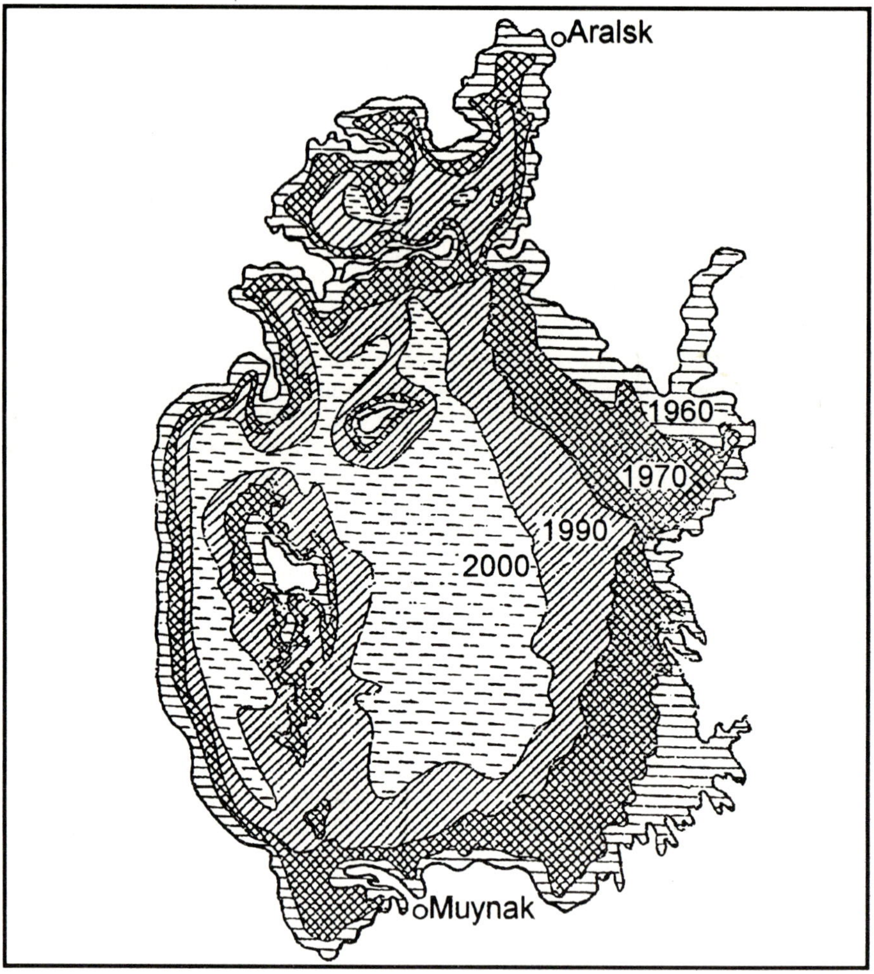

Fig. 21.1. Zones of drainage of the Aral Sea (1960 - 1970 - 1990 - 2000)

negatively on the natural environment. As a result of this, in years of little precipitation the inflow of water into the Aral Sea almost stopped. At present the water horizon has dropped by 16.8 m in comparison to 1965, the volume of water reduced by 65%, the water area shrank from 66,000-33,000 km^2, and salinity increased from 10-30 g/l. The area of dried-out seabed exceeded 3 million ha (Fig. 21.1.).

Floor sediments on a dried part of the Aral Sea bed within a one-meter layer are represented by light and medium loams, loamy sands, and sands underlain by heavy loams and clays. The prevalence of coarse-textured soils facilitates activization of wind erosion processes and the formation of eolian forms of relief.

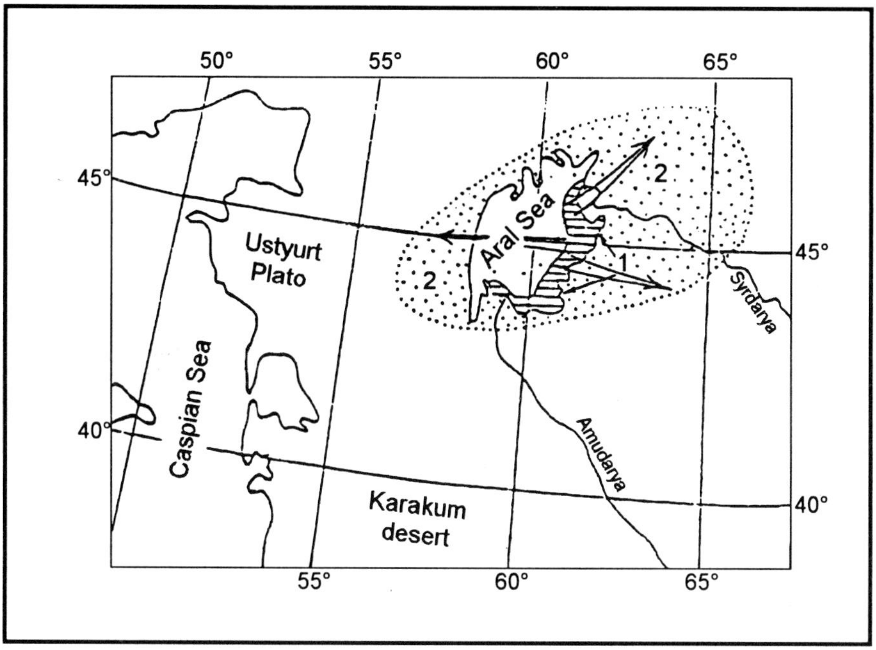

Fig. 21.2. Area of blow out **(1)** from dried eastern coast of the Aral Sea; zone of dust transfer and sedimentation in the Aral basin **(2)**

Beginning from 1975 space surveys registered significant dust drifts from the eastern coast of the Aral Sea. The length of dust trails registered from artificial satellites is usually 200-300 km. The secondary transfer of salt aerosols settled on the earth's surface significantly increases the area affected by a negative impact of the dried-out seabed on the natural environment.

In the Aral Sea basin, up to ten, powerful dust drifts are recorded every year. They are clearly visible from satellites. Analysis of photographs has shown that in 50% of cases, a dust drift went in a south-western direction to the Amudarya delta. In 25% of cases dust trails move to the west, to the Ustyurt plateau (Fig.21.2.). In the Amudarya delta on an area of 10,000 km^2 about 90-100,000 t (90-100 kg/ha) of dust material settle down on the surface every year; in the Ustyurt plateau region on an area of 13,000 km^2 about 40-50,000 t (31-39 kg/ha). The whole mass of sand driven from the dried-out part of the Aral Sea bed makes up 5-8 million tons a year. The rate of soil denudation in the exposed part of the seabed is equal, on the average 2 mm/year, according to observation data recorded over many years.

A process of dust-salt transfer in the Aral Sea basin has been going on for more than 30 years, and that is why responses of ecosystems in regions subject to these impacts are being felt acutely now. From the beginning of the 70s a steady decline of yields of pasture vegetation has been registered. The number of plants valuable in feeding terms is diminishing which, in its turn, leads to the formation of new sources of wind erosion, and the lowering of the natural potential of pastures and irrigated lands (Babaev 1995).

The Aral Sea has lost its significance as a fishery completely, navigation has stopped, and habitat conditions are progressively worsening. Impoverishment of flora and fauna has led to losses in cattle breeding and hunting. Fertile lands in the Amudarya and Syrdarya deltas have degraded in an area of about one million ha. Repeated use of return flow disposed in great quantities in rivers caused significant deterioration of water quality in major rivers. The greater part of water resources in the lower reaches of rivers are unsuitable for drinking and limitedly suitable for irrigation. The medico-biological situation in the Aral region is characterized by a high level of gastro-infectious and cardio-vascular diseases, as well as diseases of respiratory and digestion organs.

The impact of negative ecological factors on living conditions and major domains of the national economy has intensified, and so the disaster zone in the Aral Sea basin is widening (Table 21.1.).

The lowering of the sea level and desertification associated with this greatly affected the territories of river deltas. The reasons for disturbances of the ecological situation on the remaining territory of the Aral basin should be sought mostly in deterioration of the river water quality in the Amudarya and Syrdarya, and their tributaries as a result of disposal into them of untreated industrial, sanitary-domestic wastewaters and polluted drainage waters. This is combined with the slow pace of the reconstruction of irrigation systems and construction of drainage systems.

In such situation there is little hope of saving the Aral Sea, and still less hope of coordinating actions on the improvement of the situation in the Aral region.

Although the Aral problem is in the center of attention of the world community, apart from multiple meetings and publications no concrete actions have been taken so far on consolidation of efforts of international organizations and CIS countries to stop further development of negative, ecological consequences. It is necessary to prepare as soon as possible a program of coordinated practical actions to stabilize the eco-geographical situation in the Aral Sea basin, and to continue scientific researches.

Both prevention of desertification and control of desertification include a complex of social, economic and engineering measures making up a single plan of socio-economic development of any natural, territorial complex. This should be especially stressed because quite often in practical activities, attention is drawn only to engineering or technological aspects of problems, while forgetting about ecological and socio-economic aspects.

Aggravation of the ecological situation in the Aral region started from the end of the 1960s when implementation of large-scale projects of land irrigation on the basis of water resources of the Amudarya and Syrdarya rivers began. A catastrophic shrinking of the Aral Sea area and growing salinity of its waters, losses of biological resources and deterioration of people's health: all these are characteristic features of the region. In the late 1990s, the sea level has dropped by 17 m, the area of the sea has shrunk by 65%, while water salinity has grown to 35%. The dried-out seabed of the Aral Sea became a major source of salt accumulation and

Table 21.1. Total area of land desertification in the Aral Sea basin (Babaev *et. al.* 1996)

Degradation type	State of desertification						Total	
	low		moderate		high			
	km^2	%	km^2	%	km^2	%	km^2	%
Degradation of the vegetation cover	750,954	53.	307,957	21.9	23,704	1.6	1,082,615	77.0
Sand deflation	14,677	1.	2,140	0.2	3,970	0.3	20,787	1.5
Water erosion of soils	53,009	3.	29,569	2.1	—		82,578	5.9
Salinization of irrigated lands	12,959	0.	105,095	7.4	11,125	0.8	126,179	9.1
Soil salinization caused by lowering of the Aral Sea level	6,155	0.	4,027	0.3	39,055	2.9	49,197	3.6
Technogenic desertification	—		20,208	1.4	14,296	1.0	34,504	2.4
Waterlogging of pastures	—		5,360	0.4	1,620	0.1	6,980	0.5
TOTAL:	837,714	59.	474,356	33.7	9,700	6.7	1,405,840	100

salt-dust storms, too. Every year more than 65 million tons of toxic, saline dust are driven over distances of more than 500 km. Soil fertility is diminishing, pastures are degrading. The waters of the Amudarya and Syrdarya rivers are polluted with toxic chemical pesticides, harmful components of heavy metals, and carcinogenic substances.

Mortality rates of infants and mothers have sharply increased. More and more often physicians diagnose major illnesses and epidemics caused by water-born infections.

Apart from the territory of the Aral region proper, the desertification process affected some regions of the Aktyubinsk Region in Kazakhstan, the Bukhara Region in Uzbekistan, and the Lebapsky velajate in Turkmenistan. In 1989 the Government of the former USSR stated that the situation in the Aral Sea basin was beyond control, and the Aral region was declared an ecological disaster zone. Resolutions adopted by the former USSR Parliament outlined a system of measures to cope with social, economic and ecological problems of the Aral region that should be implemented by Soviet organizations (Glazovski 1990). In order to formulate and implement rational measures aimed at restoration of the ecological equilibrium in the Aral region, the Parliament resolution envisaged the establishment of a permanent governmental commission. This was charged with the task, together with the USSR Academy of Sciences, foreign scientists and specialists, of organizing a competition for development of a project, '*Concepts of preservation and revival of the Aral Sea, normalizing the ecological, sanitary-hygienic, medico-biological and socio-economic situation in the Aral region*'. 219 projects were submitted, of which 7 as containing interesting ideas were selected. They urgently prepared a project report which gave an objective analysis of the ecological situation in the Aral Sea basin.

21.1.1 International Programs

Major provisions of this concept were carefully studied by a group of international experts. The document stressed that study of the Aral problem, as is typical of many arid regions of the world, should become a model for future strategies of management of water and land resources, and the control of interaction of natural and anthropogenic factors. Proceeding from this assumption, priority was given to the UNEP Program.

Apart from a scientific plan to protect the Aral, another document was prepared '*Plan of an Integrated Use and Protection of Water and Land Resources of the Aral Sea Basin*'. It pursued the aim of developing water management and nature conservation measures with a view to ensuring for the population of the region a guaranteed drinking water supply, improvement of water quality and the sanitary-epidemiological and ecological situation, and satisfying the water demands of the developing economy's. This complex of actions was developed relying upon principles of a rational use of water and land resources and attainment of economic efficiency without detrimental consequences for the natural environment.

In early 1990 in Nairobi under the auspices of the UNEP, the international meeting was held that adopted the document '*Cooperation in Preparation of the Plan of Actions on Protection of the Aral Sea*'. This document envisaged preparation of a general plan of action, a methodology of a single approach to solution of this ecological problem, organization of monitoring, etc. that should form the basis of the international program for the stabilization of the ecological equilibrium in this region.

In April 1990 on the initiative of the Kazakh Association for the UN, Almaty was the venue of the international meeting '*How to Save the Aral*' where it was stated that the Aral crisis is a worldwide ecological disaster and it cannot be overcome by efforts of one country. The meeting stressed that it is necessary to take action as soon as possible aimed at straightening out interwoven factors: the ecological, sanitary-hygienic, medico-biological, socio-economic situation; improvement of living standards and health of the local population through the supply of good quality drinking water and foodstuffs; and the ensuring of normal living conditions for people in extreme conditions.

The disintegration of the former Soviet Union in 1991 and difficulties with financing of this project prevented the continuation of these efforts at the previous volume and pace. But still investigations and practical work along these lines are being continued in countries of the Aral region: Kazakhstan, Uzbekistan and Turkmenistan with the participation of Russia. In early 1994 in Nukus the summit of the Aral states took place, and it underlined the necessity of resolving the burning, ecological and socio-economic problems more effectively in the interests of the growing population of the Aral region. A number of inter-state, structural subdivisions were organized that are called to cope with these problems. Research institutions, design and production organization in the countries of the Aral basin keep the problem in the center of their plans and programs.

At present wide-scale, international researches are carried out into the plight of the Aral. UNEP and UNESCO demonstrate their interest in this problem. The efforts of UNEP and the World Bank to maintain an ecologically steady situation in the Aral region deserve special attention. UNEP also renders financial support to some organizations and establishments undertaking studies of the Aral problem.

In particular, the Desert Institute of Turkmenistan in agreement with UNEP and ESCAP developed a methodological basis for preparing a map of anthropogenic desertification at a scale 1:2,500,000 for the whole territory of the Aral Sea basin. At present UNESCO is undertaking implementation of the project: '*Investigation and Monitoring for Revival of the Amudarya and Syrdarya Deltas*'. Leading scientists and experts from many countries have been invited to take part in implementation of this project.

Japan, in particular the Global Research Fund, shows the greatest interest in the Aral problem. The Journal '*Problems of Desert Development*' (Ashgabat) more than once has published papers about contributions of that country to the resolution of global problems, in particular development of the world desert territories.

Between 1990-1992 The Global Infrastructure Fund of Japan not only recognized the Aral region problem as a priority issue, but initiated a campaign in favor of implementation of a program on control and management of the natural environment of the Aral region. According to Japanese business circles, the responses to the Aral crisis are exemplary for the global infrastructure that should become a framework for effective and steady development in the 21st century.

In early 1991 in Tokyo the first conference of the Global Infrastructure Fund (GIF) was held. It discussed some projects being of interest for the Japanese side. Among them is the project on improvement of natural environment conservation in the Aral Sea basin. In September 1992 the GIF together with the UN University convened the International Symposium on Problems of the Aral Sea in order to familiarize the wider Japanese public and business circles with the scale of this ecological catastrophe, and to collect enough finance for implementation of this project. Representatives from Kazakstan, Tadjikistan, Turkmenistan, and Russia, as well as UNEP and the major corporations and banks of Japan were invited to this Symposium. The Resolution adopted there state that it is necessary to set up an International Committee that would coordinate efforts of the world community on the resolution of the Aral problem. Materials of this Symposium were published in the Journal *"Problems of Desert Development"*, No. 3, 1993.

The U.S. scientific community learnt about the Aral problems, first of all, from papers of Professor Miklin, Westerm Michigan University. Apart from some papers and scientific reports, in 1991 Prof. Miklin published the monograph '*The Crisis of Water Resources Management in Soviet Central Asia*' summarizing researches of the author over many years on this problem.

In mid-1990 the University of Indiana (USA) held the International Conference on the subject '*Aral Crisis: Ecological Problems of Central Asia*' at which Soviet and U.S. scientists shared their experiences on coping with similar problems. In October 1992 at the Villanova University (Pensylvania), the Seminar '*Central Asia: Its Strategic Significance and Prospects*' was held, and there too the situation in the Aral basin was discussed.

International researches in the Aral region go on. However, this problem is so complicated and wide-scale, that it requires further extension of research works, their better coordination and major investments. In order to avoid duplication of efforts, it is necessary to organize a centralized collection and dissemination of information about investigations on the Aral Sea among scientists and experts.

21.1.2 Recommendations for the Aral

Taking into consideration that the Aral ecological crisis is far from being episodic, but will widen and deepen for a long period of time it is necessary to find solutions not only for the short-term, but also for problems than can arise in the distant future. In our opinion at present scientists and experts should consider such major strategic tasks as:

- The study of regularities that guide formation and development of desert ecosystems both in natural and anthropogenic conditions (simulation of natural processes applying computers and remote sensing methods).
- Ecological and economic assessments of desertification processes, and preparation of forecast maps.
- Desertification monitoring; development of methods for multipurpose monitoring; setting up of a coordinating monitoring center and a databank on desertification.
- The development of measures to prevent desertification processes; estimation of admissible anthropogenic loads on ecosystems of arid lands under different methods of their development; preparation of methods for planning rational management of land and water resources at various management levels; development of a methodology for accomplishment of recommendations on rational use of arid lands, including socio-economic, medico-biological and legal aspects; studies of the effect of social, economic and organizational factors on desertification processes and measures to control them; development of measures to attract people to active participation in desertification control.
- The improvement of measures to control desertification; development of agricultural practices for rainfed and irrigated conditions aimed at accumulation, storage and rational use of moisture; improvement of the methods for restoration and improvement of biological productivity of arid lands; improvement of degraded pastures; fixation and overgrowing of sands; restoration of populations; application of wastewaters in agriculture and others.

Only through a multipurpose approach to studies of natural conditions and socio-economic factors, will it become possible to create a complete model of the present state of arid landscapes in order to cope more accurately with technological and forecast problems connected with development and rational use of natural resources in the Aral Sea basin.

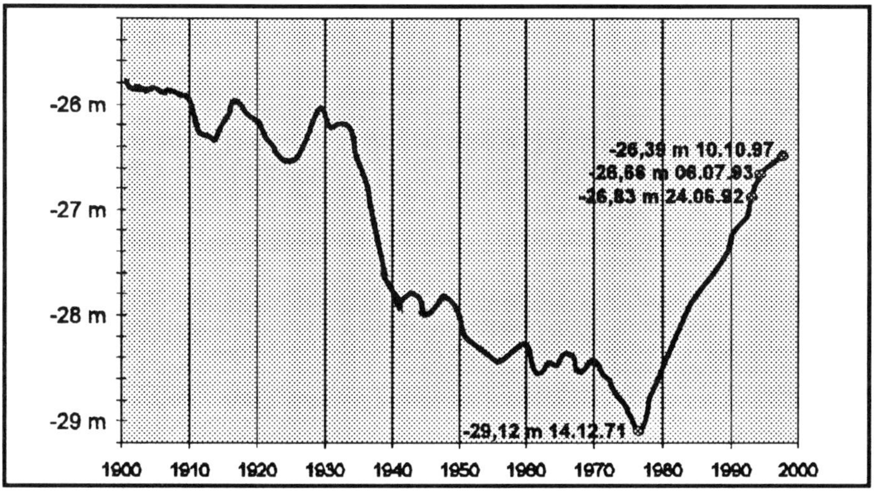

Fig. 21.3. Changes of the Caspian Sea level [m below sea level]

21.2 Caspian Sea Problems

The Caspian Sea is the largest, inland, drainless, saline water basin in the world. It stretches for 1200 km from the north to the south with an average width of 320 km. It washes the coasts of five states and connects Europe with the Near East (the coastline is about 7,000 km long). The total, water area of the sea in 1929 was 422,000 km^2, in 1969 it was 371,000 km^2 and in 1995 412,000 km^2. The Caspian Sea level is characterized by high instability: between 1830-1930 both a rise and a sharp drop of the Caspian Sea level were registered, whereas from 1931 to 1978 it dropped to 29.12 m (Fig. 21.3.). From 1977 the sea level started sharply rising again. Thus, from 1980 to October 1997 the Caspian Sea level became nearly 2.5 meters higher.

The major rivers the Volga, Ural, and Emba inflow into the Caspian Sea in its northern part. The total flow of these rivers is about 88% of the total annual river flow. The Sulak, Terek, Kura and numerous small rivers inflow into the western part of the sea and they account for only 7% of the total annual flow. The remaining 5% is contributed by rivers of the Iranian coast (Gorgan, Herad, and Sefidrud). On the eastern coast there are no permanent waterstreams.

Up to 1980, in the Caspian Sea there were about 50 small islands with a total area of 350 km^2. The largest of them are Kulaly, Tyuleny, Chechen, Artem, Zhiloy, Ogurchinsky and others. The area of islands depends on the sea level fluctuations. At present due to the rise in sea level, the areas of many islands have shrunk, and some of them have even disappeared.

The seabed relief is divided into northern, middle and southern parts. The Northern Caspian (an area of 80,000 km²) is shallow. The depth here is not more than 4-8 m. The seabed relief is a slightly rolling, accumulative plain with some shoals and islands. The Middle Caspian (an area of about 140,000 km²) is deeper. On a gentle slope, relics of ancient river valleys are found. In the south a depression of the Middle Caspian is divided from the Southern Caspian by the Apsheron structural threshold where some shoals and islands are found. The depression in the Southern Caspian, with a maximum depth of 1,025 m, covers one-third of the sea area. The bottom of this depression is a flat, abyssal plain in the northern part of which some underwater ridges running in a north-western - south-eastern direction are met.

Shores of the sea were formed by geological and hydrodynamic processes. In the northern Caspian the shores are steeper and abrupter. Here, mostly abrasive destruction processes develop. In the Northern Caspian the shores are mostly low, flat, and accumulative. The seabed is covered by sediments brought with water from the Volga and Ural rivers. The eastern shores of the Caspian are largely accumulative, low-lying, easily inundated. The proof of this is the settlement Bautino of the Mangyshlak Region, as well as the recreational zone Heles, the coastline of the Cheleken peninsula and the Balkan bay the greater part of which is already under water.

The temperature of water and air varies highly. The average, air temperature over the whole territory of the sea in summer (June-August) is 24-26°C. In winter it varies from -10°C in the north to +12°C in the south.

In the northern and middle parts of the Caspian winds often blow. From October through April eastern winds dominate, from May through September north-western winds. The strongest winds are on the Apsheron peninsula ("Bakinsky nord" blowing more often in autumn), the eastern coast of the middle part, and the north-western region of the Northern Caspian (frequent storms).

In general, up to 200 mm of precipitation a year fall over the sea water area. In the drier eastern part 90-100 mm, in the southern subtropical 1500-1700 mm.

The sea influences the climate quite specifically. It levels out air temperature variations over the whole Caspian region, thus making the climate more moderate. It provides moisture for many areas sending to the atmosphere enormous quantities of water vapors, thereby neutralizing the effect of the Central Asian deserts on the west. The Caspian basin is a unique ecological system.

21.2.1
The Causes of the Caspian Sea Level Fluctuations

Processes taking place in the Caspian Sea basin result from a combination of geological processes during a long historical period. Investigations have shown that the whole historical period of the Caspian Sea existence is characterized by many fluctuations. The major ones occurred approximately 200-300,000 years ago, when the water level rose 115 m and inundated the whole territory of the present-day Caspian, Kurinsky, and Western Turkmen lowlands. The shores of the Caspian Sea reached to Samara in the north, to Mingechaur-Gyandzha, Stavropol in the west and to Ashgabat latitudes in the east.

We believe that the sharp rise of the Caspian Sea level today is a result of climatic, geological, anthropogenic and other processes.

Climatic Factors

As early as in the 30s, the senior academic Berg spoke about a relationship between the Caspian Sea level fluctuations and ice conditions in the Arctic seas (Berg 1980). He believed that in the period of climate warming in the Arctic, the amount of precipitation in the north of the European part of Russia, in particular in the upper reaches of the Volga and Kama, tended to decrease. This led to low flow in rivers and, as a result, a fall of the Caspian Sea level.

While not disputing the great influence of climatic factors on the behavior of the Caspian Sea level, not all scientists support Berg's hypothesis based as it is on the difference of incoming and outgoing balances of water resources. In particular, the present rise of the sea level does not match the values of the surface flow of rivers, precipitation and evaporation. Thus, according to data of the former USSR State Committee for Meteorology, in 1980 by the sum of inflow, precipitation and evaporation the sea level should have risen by 3.8 cm. In fact, the rise was only by 11 cm. In 1982, when the inflow and precipitation were below the norm and evaporation within the norm, the Caspian level should have lowered by 4 cm, but, in reality it rose by 8 cm. In 1993 the surface inflow and precipitation were lower, and evaporation higher than the norm which, according to estimations, should lead to the level lowering by 10 cm. In reality it again rose (Shilo and Krivoshei 1989). Such deviations prove that the present-day rise of the Caspian level and even more so the yearly increases of water intake from rivers for industrial, agricultural and domestic purposes should be explained by the action of other factors.

Geological Factors

Geologists draw attention to a relationship between the Caspian Sea level fluctuations and tectonic dislocations in the Earth's crust observed in this region (Babaev *et. al.* 1996, Shilo and Krivoshei 1989). These relationships should be sought in specific geological structure of the Caspian depression consisting of three basins: the Middle Mesozoic northern (platform) basin and the southern, younger and deeper basin divided by the Apsheron threshold stretching from Baku to the Krasnovodsk peninsula. This part of the sea belongs to the Alpine fold area which is distinguished by sharply differentiated, vertical dislocations. Differences in dislocations between a platform part and a basin in the Southern Caspian change in time, and can reach considerable values. This, of course, should result in a changed capacity of the whole Caspian depression and cannot but affect the sea level.

Shilo and Krivoshei explain variations of the Caspian Sea level by compression and extension of sedimentary rocks that make up the Caspian seabed (Shilo and Krivoshei 1989). They believe that the Northern Caspian was controlled by sub-meridional compressive forces from the Russian platform, while the South-Caspian basin enters into a sublatitudinal extension zone of the Caucasus-Iranian, young fold area. Such tectonic structures influence the deep-lying, sedimentary rocks of intercapillary waters. Affected by compressive stresses, the latter give their waters to the Caspian, and subject to the effect of extending forces or with a weaker compression they take waters back. Consequently, the compression stage coincides with the rise of the sea level, and the extension stage with the lowering of the sea level.

Earthquakes nearby the Caspian Sea can cause outflows of groundwaters. Earthquakes in shallow-water areas with a magnitude less than 6.5 impel groundwaters to outflow into the sea (28 km^2/month). Earthquakes with a magnitude of 6.5 cause outflow of water from the sea (35 km^2/month). During earthquakes in deep-water areas inflow and outflow double.

Anthropogenic Factors

Humankind's activities have produced enormous effects on the state of climate and the biosphere in general. Progress in science and technology led to new kinds of impacts on the natural environment. In particular, the climate of the Earth changes due to the greenhouse effect, the ozone layer is shrinking, the lower atmosphere, soils, rivers, lakes and, in general, oceans are polluted. Burning of large quantities of organic fuel leads to increased amount of carbon dioxide in the atmosphere, and as a result the emission of solar and deeper heat by the Earth's surface to outer space is delayed. Climatologists believe that by the beginning of the 21st century a temperature of an underground layer will become 1°C more and in subsequent decades 2°C more.

Therefore, supporters of the climatic hypothesis explain the Caspian Sea level rise by a global warming, because it is attended by the increased evaporation from the surface of seas and oceans and, as a result, by the increased precipitation over continents, including the European part of Russia from where the Volga, Kama and Ural rivers originate – the principal suppliers of water resources to the Caspian Sea. Consequently, the flow of the Volga with its tributaries increases which will facilitate further rise of the sea level.

It should be said that there is another opinion. Climatologists of Kazakhstan, analyzing statistics on the Caspian Sea level for the future till the year 2050 taking into consideration likely variations of climate, water use in the basins and inflow of sea water into the Kara-Bpgaz-Gol Bay, think that a further, continuous rise of the sea level is unlikely. Moreover, they predict that a rise of the sea level will stop in the nearest years. The level will stabilize, and maybe even tend to some lowering.

Possible Connection with Solar Activity

In 1945 Eigenson showed that a physical cause of warming in the Arctics and modern variations of climate in some world regions is the variation over centuries of solar activity. More than 40 years ago Appolov, Belinsky, Bitelje, Bezrukova and some other scientists pointed to a relationship between century-long fluctuations of the Caspian level and solar activity. These problems were developed in thematic works undertaken by the Desert Institute together with the Institute of Earth Magnetism of the Russian Academy of Sciences.

Movement of the Earth's Crust

Other scientists believe the cause is connected with oscillatory movements of the Earth's crust. The catastrophic events that are witnessed today on the Caspian can be related to different natural phenomena, including processes in space.

In our opinion one of the main causes of fluctuations of the Caspian Sea level is related to geological processes underway in the deeper layers of the Earth's crust, in particular to long-term oscillations as a result of which some parts of the Caspian Sea bed e.g. in the southern part of the Caspian basin.

Century-long oscillation movements are a unique phenomenon in geology discovered by leading figures of geological science A.P. Karpinsky, N.A. Golovkinsky, J.K. Gilbert, G. Shtille, V.V. Belousov, A.L. Yanshin and others.

M.V. Lomonosov described these oscillation movements as "non-sensitive and long-term lowerings and uprisings" of the Earth's surface. These oscillatory movements, unlike sharp tectonic disclocations and earthquakes do not occur rapidly but rather slowly and invisibly. It is with these movements that slow and invisible risings and lowerings in the lithosphere are connected. They occur even now, and will go on always until the planet Earth exists; this could be expressed as the Earth breathing.

We made a comparative analysis of changes of the depth of the Caspian Sea bed by maps prepared in different years in the 20th century (Table 21.2.).

According to the physical map of the Caspian Sea, the greatest depth in the Southern Caspian basin is 945 m, of the Middle Caspian 768 m. According to the '*Greater USSR Atlas*' these figures are 980 m and 790 m, respectively. The navigation directions of the Caspian Sea states that in 1967 the deepest point in the Southern Caspian basin was 1020 m, and in 1975 1025 m (42°00', 50°00', 0); in the Middle Caspian 790 m and 788 m, respectively (39°00', 49°10', 0).

Therefore, analysis of cartographic materials proved that between the 1930s and 1980s, the floor of the Southern Caspian basin sagged by approximately 80 m. Probably this recession of Caspian Sea waters in the mid-20th century is connected with these processes of lowering of this enormous depression. It cannot be excluded that since the 1980s a gradual and hardly perceptible rise of the bottom of some basins in the Southern Caspian depression has begun. The modern rise of the Caspian Sea level is connected with this very process.

Table 21.2. Comparative analysis of changes of the depth of the Caspian Sea bed (1930-1994)

Map	Year of Publishing	Maximum depth [m]	
		Southern Caspian	Middle Caspian
Greater World Atlas, sheets 88-89	1937	945	768
Greater USSR Atlas, sheet 37	1946	980	790
Navigational Directions of the Caspian Sea, p. 17	1967	1020	790
Navigational Directions of the Caspian Sea, p. 15	1975	1025	788
Geological Map of the USSR	1984	1025	788

An indirect proof of a rising tendency of the Caspian Sea bed may also be the activization in the recent years of mud volcanos locating along the coastal zone of the Southern Caspian basin. Results of expeditions have shown that in the recent decade most active are the mud volcanos Ak-Patluk, Porsu, Gainak, Chekishlyar in the Esengulisky etrape, and Rozovyi Porsychel on the Cheleken island.

Valid evidence of the rate, degree and nature of oscillatory movements can be obtained in the course of stationary, instrumental observations. A continuous rise of the sea level threatens aggravation of the ecological situation in its basin. First of all, this is a threat of flooding of industrial and civil facilities, roads and agricultural lands in the coastal zone in Turkmenistan, Azerbaijan, Russia, Kazakhstan and Iran.

21.2.2 Areas Threatened with Flooding

In Turkmenistan the low-lying settlement Cheleken is already under water, as well as the recreational zone Heles, port Ala-Tepe, Aladja and partially the recreational zone Awaza. Communications of the Turkmenbashi oil refining plant, objects in settlement Ufra, settlements Esenguly and Chekishlyar are threatened with flooding. According to the Ministry of Ecology and Bioresources of Kazakhstan, 43 oil and gas fields are in the flooding zone.

In Russia coastal lands in Daghestan, Kalmykia and the Astrakhan Region with an area of 1,655,000 ha with approximately 50 towns and settlements with a population more than 100,000 are in the zone of nearest flooding.

21.2.3 Deterioration of Water Quality

Deterioration of the Caspian water quality as a result of disposal there of polluted flows of the Volga, Ural, Kura, Terek rivers, industrial and domestic wastewaters, as well as accidents during drilling and production of oil and gas causes great damage to biological resources of the sea.

At present the average, annual concentrations of petroleum products and phenols in the area of the northern and eastern coast of the Caspian exceed MAC 4-6 times, while levels of sea pollution for petroleum products and phenols alone near the Azerbaijan coast by 10-15 times. Petroleum products disposed into the sea severely affect the quality of caviar, larvae, fish fries, destroy spawning grounds, and eliminate a forage base. For the years 1983-1989 the sturgeon population in the Northern Caspian was reduced by 7 million, and the figures for the area near the Azerbaijan coast are still worse.

The Caspian Sea is the final basin for flow of many anthropogenic components. Every year more than 2.5 tons of heavy metals, 62-146,000 tons of petroleum products, and 4.5 million tons of biogenic pollutants are disposed of in the delta of the Volga, Terek and Sulak rivers. From nearby Baku alone about 300 million m^3 of untreated wastewaters are disposed into the sea. More than 100,000 m^3 of toxic substances a day are emitted by industrial enterprises in Sumgait, and they get into the Caspian Sea.

Poor technologies, non-integrated and often barbaric use of natural resources, wasteful economic mechanisms and unrenewable management of the environment: these are the causes why vast territories, especially gulfs and bays on the Caspian, turned into a dead zone. Unfortunately, the boundaries are widening and in the face of maintaining the existing technogenic load, the situation might soon become irreversible.

At present it is naturally impossible to predict what will be the sea level in the nearest decade. But by analogy with discussed periods, one cannot exclude the sea level rising by another 2.5-3 m, and this level may be maintained for a long time.

21.2.4 Recommendations

In order to save the towns, settlements, ecological objects, population around the Caspian Sea, it will be necessary in the next decades to stop any geophysical explosions, any artificial seaquakes at the bottom. Work on deep surveying, prospecting and drilling for oil and gas in the water area of the Caspian Sea should be suspended for some time.
It is necessary to diminish catches of sprat and small-net fish. For more than 30 years there were mass catches of sprat with the help of powerful technogenic means (fat-flour factories) representing huge floating enterprises that "devoured" millions of tons of small-net fish. Meanwhile the people living in coastal regions and suffering from a deficit of foodstuffs were not allowed to go fishing. This contributed to some extent to the growth of social strain.

In order to avoid the risk of flooding, the major structures (industrial plants and settlements) should be constructed in areas in which elevations are not lower than minus 23 m, and composed of pre-Quarternary deposits (as in Aktau, the upper settlement of Cheleken and elsewhere) and not more recent, accumulation formations. Construction of industrial and civil objects should take into account a geological structure and engineering-geological properties of rocks in a region.

In the very near future, it is necessary to undertake technical engineering works, including floor dredging and construction of embankments in order to mitigate the impact for ports and anchorages caused by the rise of the sea level. In the scientific domain it is necessary to initiate in-depth investigations of natural resources and specific ecological features of the Caspian basin in order to set up the Inter-State Multipurpose Institute on the Caspian Problems with its basis in the existing stations and posts.

For an optimal solution of the problems of the Aral and Caspian Seas, it is very important to organize international cooperation in the exchange of information on efforts undertaken by the states located in the basins of these seas, international and national organizations. Such cooperation will facilitate coordination of actions aimed at improvement of the quality of the natural environment in the regions, and implementation of respective national, regional and global ecological programs.

Regional cooperation aimed at improvement of the natural environment and removal of social strain in the basins of the Aral and Caspian Seas is an inseparable part of the overall problem: the harmonization of the system "nature-society".

References

Babaev AG (1995) Problems of desert development. Ylym, Ashgabat.

Babaev AG, Amanniyazov KN, Muradov, CO (1996) Fluctuation of the Caspian Sea level and it's ecological consequences. Probliemy osvoieniia ustyn: 1996, No. 6.

Berg LS Caspian Sea level for histric time. Selected works. Moscow.

Glaszovki NF (1990) Aral disaster. Nauka, Moscow.

Shilo NA, Krivoshei MI (1989) Connection of fluctuations of Caspian Sea level. Vestn. AS USSR, 1989, No. 6.

22. The Repetek International Biosphere Reserve

J.V. Rakhmanov, S.V. Veyisov, V.Y. Kuzmenko

ABSTRACT: The Repetek reserve is unique. Founded as a desert research station in 1912, in 1927 it acquired the status of a nature reserve. In 1979 UNESCO recognized it as an international biosphere reserve. Climate and relief are described along with the richly variegated flora and fauna. It is a base for significant international reserve. The long-established museum educates the public about ecological themes, and the whole area plays a practical role in the conservation of species.

In 1979 UNESCO took the decision to assign to the Repetek Sandy Desert Nature Reserve the status of a biosphere reserve. The extraordinary landscape, unique vegetation, minimum anthropogenic impact on development of the natural complexes, combined with venerable, scientific traditions and great experience in studies of ecosystems of the Karakum sandy desert were the factors that made it possible to recognize the Repetek reserve as a biosphere reserve (Babaev 1989).

22.1 Location

The Repetek Biosphere Reserve in Turkmenistan is located in the Eastern Karakum, 70 km to the south-west of Chardjou city. The total area of the nature reserve is 34,600 ha. The whole territory of the nature reserve is covered by a sandy desert. Coordinates of the nature reserve are as follows: from 38°27'2" to 38°42'0" of the north latitude and from 63°05'8" to 63°23'0" of north longitude. The altitude of the nature reserve over the sea level is 185-200 m. In 1959 the Repetek Biosphere Nature Reserve was incorporated in a system of the Turkmenistan Academy of Sciences as a large, structural unit of the Order of the Red Banner Institute of Deserts.

22.2 History

The name "Repetek" most probably came from the Arab "*Rubshatakh*" or "*Rafatak*" is translated as "narrow space with fresh water". For the first time it was mentioned in historical documents of the 19th century, when one of the roads to Bukhara "Repetek-yol" passed through there.

In the second half of the 19th century Russia started construction of the railway Krasnovodsk-Chardjou in the trans-caspian region. However, they faced serious difficulties here connected with drifting sands. The threat of sand drifts remained during operation of this railway, too. So the administration of this railway applied to the Russian Geographical Society with a request to help them with protection of the railway from moving sands. The Russian Geographical Society supported this request, and in 1910 they set up a commission to study drifting sands. The commission made a recommendation of the necessity of establishing a scientific station in a sandy desert nearby the railway station Repetek where at the end of the 19th century Paletsky had organized an afforestation nursery. In the spring of 1912 the Repetek sandy desert station was opened with Dubyansky as its head.

In the period from 1912 to 1918 the station initiated the first meteorological observations in the Karakum desert, and developed methods for protection of the railway from sand drifts. In the second half of the 1920s expeditions were sent to the Eastern Karakum for geobotanical inspection.

On 27 October 1927 the Repetek State Nature Reserve was established on the basis of the sandy desert station. Among its goals were the preservation and study of flora and fauna on the undisturbed desert territory.

In the 1930s, the nature reserve undertook a large volume of work on growing agricultural crops in a water-deficit conditions, studies of physiology of psammophytes, flora and vegetation. In the 1950s, they completed research work on the introduction and growing of trees in the nature preserve, near railway stations and crossing points, and the construction of pits on pastures. In the 1960s, the nature preserve carried out work as part of the International Biological Program, and determined the biological productivity of desert communities. Beginning from the 1970s they undertook integrated studies of ecosystems of the Karakum sandy desert. After the Repetek Nature Reserve acquired the status of a biosphere reserve, they started developing the foundations of the land-based monitoring of natural components in sandy deserts. In the nature reserve they regularly organize international training courses for specialists in nature conservation from developing countries of Asia, Africa and Latin America (Veyisov *et. al.* 1996).

22.3 Geomorphology

The territory of the reserve is an alluvial plain composed of the Lower and Middle Quarternary deposits of the Pre-Amudarya consisting largely of schist-like gray sands of different granulometry interbedding with clays and gypsum layers. Alluvial deposits are overlain everywhere with migrated eolian sands whose thickness varies from 0-30 m.

The sand relief is highly variable. The greater part of the territory is covered by hillock sands with an area of 18,600 ha. About 14,000 ha are covered by barkhan sands. On the territory of the nature reserve one can find valley-like depressions with thickets of, *Holoxylon persicum*. Their total area is about 2,000 ha (Plate 22.1.).

Groundwaters, largely fresh, occur at a depth of about 30 m under mobile barkhan sands. Their salinity is 0.5-1.0 g/l. Inter-ridge hollows overgrown with sea sedge are composed of typical sandy soils. Saline groundwaters (5-10 g/l) occur at a depth of 5- m. Valley-like depressions are distinguished by well-formed soils of low and medium salinity overgrown with thickets of *Holoxylon persicum*. Slightly saline groundwaters (1-3 g/l) occur at a depth of 5-9 m.

22.4 Climate

The territory of the Repetek Nature Reserve has a climate of continental, subtropical deserts. The annual amount of incident solar radiation here is about 160 kcal/cm². The average air temperature is +16°C. Observations over many decades have shown that average monthly temperatures here are always positive. The hottest month is July when the average, monthly temperature is +31°C. The highest air temperature (50.3°C) was registered in Repetek on 29 July 1983 – the hottest for the Asian continent. On such days temperatures of a bare sand surface can reach as high as +77°C. Negative temperatures are recorded from December through February, while a frost-free period lasts for about 300 days. The lowest average, monthly temperature is observed in January (+1.7°C). The absolute minimum is -31°C. Air temperatures on average reaches 20-25°C.

The average, annual sum of precipitation falling mostly in the form of rain is equal to 113.6 mm. Their distribution over seasons is very uneven: in winter 37%, in spring 49%, in summer 2% and in autumn 12% of the annual amount of precipitation. The ratio of precipitation in the dryest year 1917 (24.3 mm) and the wettest year 1969 (229.5 mm) is almost 1:20. Against a background of sharp variations of a temperature and uneven precipitation, very large decreases of relative air humidity are observed within a year varying from 5 to 80%.

The average winter in the nature reserve starts usually in early December, spring in mid-February, summer in late April, and autumn in late September. The winter is characterized by average daily temperatures below +5°C, summer above +20°C. In winter, frosts are often replaced by thawing periods. Quite often in January, sea sedge and other grasses can be green. In spring during a very short period the desert gets covered by colorful carpets of ephemeral vegetation. Summer in the Karakum desert is dry and hot. At the beginning of summer all *Calligonum rubens, C. setosum* plants bear fruits. In July-August a day air temperature in a shadow is usually higher than 40°C. Only during bright nights due to intensive radiation cooling, can one enjoy freshness of air. Autumn in the nature reserve is associated with the fruit-bearing of *Haloxylon*. Bright, calm weather with cool nights reigns here at this time (Gumin *et. al.* 1991).

22.5 Flora and Vegetation

Flora in the nature reserve has about 120 varieties of aboriginal flower plants, 8 varieties of fungi, 4 kinds of mosses and 4 kinds of lichens, 68 varies of soil water grasses and 145 kinds of micromycetes only on overground organs of plants. Seven families form the basis of flowering plants, including a typically desert, goosefoot family represented by 22 varieties (18.3%); then come the thistle and cruciferae families 16 varieties of each, (or 13.3% of each); cereals 9 varieties (7.5%), legumes and borage 8 varieties of each (or 6.7% of each), then the buckwheat family 7 varieties (or 5.8%). Among life forms annual grasses dominate with 76 varieties (63.3%) perennial grasses count 23 varieties (or 10.2%), shrubs 11 varieties (or 9.2%), trees 4 varieties of each (or 3.3% of each), semishrubs 2 varieties or 1.7%.

Flora in the nature reserve is characterized by a high specialization of plants by xerophylic properties (adaptability to a live in conditions of a moisture deficit) and psammophylic properties (adaptability to a life among drifting sands), and a great number of varieties endemic to the Karakum and Kyzylkum (37%). Among endemic varieties most typical are *Calligonum spec., Stipagrostis spec.*

Dominating the vegetation cover in the nature reserve are communities of plants of a psammogenic succession. On barkhan sands *Erkekselin kandym* grows; on barkhan hillocky and hillocky non-ramped sands one often finds *Erkekselin kandym* with *Haloxylon aphyllum*, and on hillocky ramped sands *Haloxylon aphyllum* with *Carex physodes*. In the valley-like depressions, *Haloxylon persicum* vegetation connected with groundwaters are widespread. These "forests of a desert" cover about 5% of the nature reserve territory. More than 100 varieties of plants among which grasses dominate are found here. The Repetek nature reserve is one of the few places in the Karakum where these forests are preserved in their natural state. The greatest area in the nature reserve is covered by *Haloxylon aphyllum*.

The sandy desert is characterized by the prevalence of non-leafy varieties, the early start of a generative phase and a short life period. Green sprouts, e.g. *Ha-*

loxylon spec., perform a photosynthetic function. Trees and shrubs in the nature reserve start flowering mostly in the 4th-6th year, and live for more than 30-40 years.

In conditions of poor mineral nutrition and inadequate moisture supply the trees and shrubs in a desert must have well-developed root systems. Vertically they can penetrate to a depth of 7-8 m and more, and horizontally they can spread to 25-30 m. This is the reason why even phytocenoses of barkhan sands where an over-ground horizon is very thinned the closeness of root systems of plants is as high as 50%.

Dominating in a grass horizon are ephemerals and ephemeroids adapted to summer droughts. In a dry period they do not vegetate at all, they survive it in a state of dormant seeds (ephemerals) or dormant subsurface organs (ephemeroids). As these plants are active only in spring where air is still humid and soil is wet from rainfall, they didn't have to change their way of life as is the case with drought-resistant xerophytes. That is why ephemerals and ephemeroids do not have a "desert" appearance. They resemble by the appearance their closest relatives from zones with moderate climate.

Ephemerals can be seen in April and May in all their glory. In this period meadows covered with motley grasses in *Holoxylon persicum* growths are very colorful and dazzling. Vast meadows covered with bright-red poppies are replaced by small spots covered by yellow-white pyrethrum or rose larkspurs. Meadows are framed with slender bunches of light violet *Strigozella*. This luxurious carpet is pierced by green panicles. Already in mid-May these formerly verdant meadows have been dried-out by a fierce sun, and only brown stems remain from the former wild beauty of the spring scenery.

Among rare plants that are growing in the nature reserve it is necessary to name long-stem iris, and Turkestan rhubarb with its very large leaves and red-rose flower clusters. Many plants growing in the nature reserve are used as medicinal plants.

Many plants in the sandy desert live by absorbing moisture that is contained in sand. Only *Holoxylon persicums*, the most productive community in a desert, intensively use low-saline groundwaters (Plate 22.2.). Under favorable conditions, roots of *Holoxylon aphyllum* also reach groundwaters. Productivity of raw over-ground mass of plants varies from 250 quin/ha for *Holoxylon persicum* to 1 quin/ha for *Calligonum rubens, C. setosum* on barkhan sands. The mass of green sprouts of sea sedge, the most valuable pasture plant, is 1-2 quin/ha in the Kara-kum.

22.6 Fauna

Very diverse and specific is the animal world of the nature reserve. Even in bar-khan sands an attentive observer will see inhabitants of this at first sight, lifeless terrain. Everywhere one can see traces of jerboas and beetles. On a high barkhan, a lizzard stands as if frozen turning its head to a wind. A disturbed jay flying from one bush to another cries angrily. In the morning and in the evening, one can easily detect vertical holes of tarantulas which they close during a hot day.

The Repetek nature reserve is inhabited by 26 species of mammals (83.9% of the total number of the Karakum species). Prevailing among them are rodents, especially sandworms and jerboas. Hares are often met. Rarely met here are porcupines. Of the insectivorous mammals most widespread are long-eared hedgehogs, of predatory mammals the foxes. Weasels, sand and spotted cats, Central Asian gazelles are among the rare animals one can find here.

The most numerous group of vertebrate animals in the nature reserve is avifauna including 198 species (91% of the total number of the Karakum species), of which there are 15 species of settled, 18 of migratory-nesting, 25 of migratory-hybernating, 122 of migratory, 3 of nomadic and 16 of passage birds. Among settled birds there are saxaul and sand sparrows, Bukhara tomtit, saxaul jay and others. The greatest population of sand sparrows known only in several places in the Karakum and Southern Kyzylkum, live in the Repetek nature reserve; here 123 pairs of them are found. Saxaul jay is an endemic species of the Karakum desert.

Among vertebrate there are many rare and endangered species that are put in the Red Book (conservation monitor) of Turkmenistan. These are Central Asian gazelle, golden eagle, sand sparrow, Central Asian cobra, caracal, gray monitor lizzard, short-toed eagle.

Twenty-two species of reptiles (95.6% of the total number of the Karakum species) are found in the Repetek nature reserve: Central Asian turtle, grey monitor lizzard and four other species of lizzards, nine species of snakes to name but a few. Some herpetofauna have the status of rare and endangered species.

The invertebrate fauna is very diverse in the nature reserve. Now it has been established that 1231 species and subspecies live here: this is 25% of the estimated number of species which are thought to be resident in the Karakum. Arachnid and insects dominate. Of the arachnid in the Repetek nature reserve tarantula, 5 species of scorpions and more than 10 species of phalanx are to be found. Of the *Coleoptera*: 494 beetles species, *Lepidoptera*: 188 butterfly species, *Hymenoptera*: 157 species. They make up the basis of entomofauna of Repetek. Among the invertebrate too there are many rare species.

22.7 Scientific Activities in the Nature Reserve

In Repetek scientific investigations are carried out the whole year round. For many years the Repetek station took a leading place in desert studies in Asia and North Africa. Very advanced, scientific ideas in physical geography, botany, and zoology in the desert zone were first developed in Repetek. Here they first studied regularities in the development of a relief of barkhan sands, groundwater formation, specific features of climatic and microclimatic factors in a sandy desert, a moisture regime of soils, studied anatomy and physiology of plants, dynamics of the vegetation cover over many years, problems of photosynthesis, biochemical cycle of matter, ecology of vertebrate and invertebrate animals, soil fauna, a structure of antomocomplexes, primary and secondary productivity, developed new methods of study of populations of plants and insects, the foundations for restoration and use of saxaul vegetation, and recommendations on protection and rational use of natural resources in a sandy desert.

For the results of long-term scientific investigations carried out in Repetek, 10 theses for a doctoral degree and more than 40 theses for a candidate degree have been defended. More than 1500 scientific papers, including 30 monographs and collections have been published.

In Repetek such well-known scientists as Dubyansky, Orlov, Gael, Leontiev, Petrov, Vasilevskaya, Semevsky, Kunin, Blagoveschensky, Znamensky, Stalmakov, Mikhelson, Gunin, Dedkov, Kaplin, Beyisov and others have worked.

From 1981 on a small team at the Repetek station have worked on the problem, "Ecological monitoring of sandy desert ecosystems". During this period for the first time in the Eastern Karakum, schematic maps of ecosystems and their present state were prepared, including distribution of *Holoxylon persicum* vegetation using data obtained from satellites and the results of field expeditions. Summaries have been published of materials concerning long-term dynamics of the population and age spectra of cenopopulations of edificators in the primary, psammogenetic succession, and materials describing a rate of this succession in a landscape of large-ridge sands. They analyzed biomorphological specific features of shrubs, and their adaptability to life on a sand substrate.

For the first time in the world, they completed a planned biogeocenological study of complexes of invertebrate animals. From 1981 onwards they have completely processed these materials, and revealed regularities in respect to complexes of invertebrate animals living in tissues of plants, on plants, in soil, and in the underground habitats of rodents. They appraised biomass and the role of vertebrate and invertebrate animals in sandy desert ecosystems, in general.

In the future it is planned to continue ecological monitoring in the Repetek nature reserve, on the adjoining territory and in the Eastern Karakum. It is designed to pay much more attention to a comparative study of similar ecosystems in natural and anthropogenic conditions in order to develop recommendations on assessment of their state, conservation and rational use.

The scientific station and the nature reserve are of great importance as a year-round station that makes it possible for CIS and foreign desert scientists to carry out continuous in-depth investigations in one of the few terrains of a sandy desert preserved in its natural state. Workers of the station and the nature reserve take part in investigations of the world deserts, in international scientific forums.

A herbarium fund of the nature reserve incorporates two sections: reference and stock. A reference herbarium section, about 600 sheets, includes plants growing on the territory of the nature reserve. A greater part of the stock herbarium (about 1,000 sheets) is collected in the Karakum and other regions of Turkmenistan, in Uzbekistan and Kazakhstan. The ealiest collections belong to Androsov, Mikhelson and are dated 1893-1903, 1909-1911. These were given as a gift from the former museum of sand stabilization in Farabe of the Chardjou Region of Turkmenistan.

The herbarium stock of the nature reserve has also about 400 sheets of plants from the southern deserts of USA. This herbarium was collected between 1916-1936 in California, Arizona, New-Mexico, Oklahoma, and Texas. This herbarium was presented to the Repetek sandy desert station and nature reserve by the U.S. National Herbarium.

A museum of nature was set up in 1972 in Repetek mostly with large, vertebrate animals. There are stuffed figures of 13 species of mammals, 62 species of birds and 21 species of reptile. Exhibits in the museum are arranged in biogroups and especially interesting for visitors is the small collection of beetles, black beetles and butterflies.

The museum also displays some archeological findings: four jugs in a well-preserved state (from the Middle Ages); three of them were obviously intended for liquid products, and one large jug for grain. Visible on some broken pieces of pottery covered with lacquers of different colors are the relief figures of a man.

The museum also shows cuts of various trees and shrubs of the nature reserve, as well as on-wall herbarium exhibits of some grass plants. It also displays a collection of fruits of desert plants.

From the early 1960s after construction of a water main, great attention in the nature reserve is given to planting trees and shrubs in the central estate. In the early 1970s more than 70 varieties of desert, fruit-bearing and decorative tress and shrubs were growing here. Now you can find in the park not only desert plants, but also some varieties of *Populus*, *Acer*, *Morus*, *Elaeagnus*, *Punica* and others. Earlier in 1947 at a distance of 200 m to the north of the central estate a turanga was planted in a trench. At present it forms a shadowed grove demonstrating a possibility of rainfed growing of tugai plants in conditions of a sandy desert.

Workers of the nature reserve give much efforts to extension of knowledge on scientific achievements in study, rational development and protection of desert territories. The Repetek nature reserve is one of the most popular places which is visited by many people. Every year it welcomes about 5,000 scientists and lovers of nature, including 200 foreign guests. Organization of excursions and lectures in the museum is one of the most important lines of activity of the station in extension of knowledge and ideas of protection and ecological education of the population.

These activities help preservation of unique and very valuable landscapes, rare and endangered species of plants and animals. They do much to change the outlook of the public and population on the monuments of nature in their native land, to educate people towards ecological thinking and to revive nature conservation activity in the region.

References

Babaev AG (1982) Anniversary of Repetek sandy-desert station. Probliemy osvoieniia pustyn. 1982, 4: p. 3-7.
Gumin PD, Darymov VY, Veyisov S (1991) Landscape characteristic of Repetek reservation. Izv. AS. TSSR. Ser. Biol. 1991, 14: p. 39-47.
Veyisov S, Kuzmenko VD, Sopyev O (1996) Repetek international biosphere reserve. Probliemy osvoieniia pustyn. 1996, 4: p. 25-30.

23. International Cooperation in Desertification Control

A.G. Babaev, C.O. Muradov, A.A. Khachaturyan

ABSTRACT: The Desert Institute has played an active and important part in the international efforts to research and combat desertification. The former USSR was one of the leaders in coordination and integration in this field working especially through UN organizations. Many seminars have been organized by the Institute in conjunction with UNEP, the World Bank and others. The Institute has acted as consultants in such countries as Mali and Peru, and taken part in projects with China, India, Israel etc. Moreover, they have prepared a number of project proposals concerning desertification and increased global synchronization which is undoubtedly needed.

The Desert Institute of Turkmenistan has been for more than 30 years one of the centers and coordinators of investigations on desert problems and desertification control.

In 1967 through a resolution of the USSR Academy of Sciences, the Desert Institute was assigned the function of the head research institution and coordinator of investigations on the problem "Integrated study and development of desert territories of Central Asia and Kazakhstan". At the same time a scientific council was established on this problem. Furthermore publication of the scientific journal *"Problems of Desert Development"* was organized: it describes the results of major scientific researches and practical experience gained concerning the study and management of desert territories. The journal was first published in Russian with 6 issues a year. From 1980 the publishers *Allerton Press* (USA) undertook publishing and distribution of the journal *"Problems of Desert Development"* in English-speaking countries.

23.1 The Emergence of the Desertification Phenomenon

The problem of desertification acquired worldwide significance through the extremely severe drought of 1968-1973 in the Sudan-Sahelian zone, of which thousands of people died, caused enormous damage to pasture cattle-breeding and rainfed farming in many developing countries of Africa.

The first UN world conference devoted to this problem was held in 1977 in Nairobi. Not only did it touch on the fates of millions of people, but also the very existence of the fragile ecosystem of the arid zone. This forum was also very important for exchange of experience amassed in many countries on development of desert territories and desertification control.

The authoritative Soviet delegation was headed by A.V. Sidorenko, the Vice-President of the USSR Academy of Sciences Academician and his deputy, A.G. Babaev, the President of the Academy of Sciences of Turkmenistan, Director of the Desert Institute, and member of the USSR Academy of Sciences. The two scientists took an active part in the work of this conference.

The report of the Soviet delegation at the conference was prepared relying upon the results researches of many years and the fifteen-year experience of the Desert Institute. For this conference the *"Mysl"* Publishers published the book by Babaev and Freikin "USSR Deserts: Yesterday, Today and Tomorrow". It summarized an enormous amount of material about deserts of the Soviet Union, mostly in Central Asia.

The constructive stand of the Soviet delegation at the conference revealed itself in the fact that after careful study of the "*Plan of Action to Combat Desertification*" which had been submitted for discussion, and after applying its wealth of experience in the development of arid territories in Central Asia, it came out with a proposal to put in the Plan recommendations concerning a combination of industrialization and urbanization with development of agriculture in the course of development of arid territories and the likely ecological consequences.

The Soviet delegation proposed to study and summarize the role of industrialization and urbanization in development, prevention and elimination of desertification processes in arid regions under various socio-economic conditions; to render all-round scientific and advisory aid to developing countries affected by desertification processes in study and planning of industrial and urban development; to prepare and publish reviews and monographs describing positive experiences of socio-economic development, and the protection and improvement of the natural environment of arid territories. Last but not least it mooted the possibility of organizing and holding a number of meetings for discussion of the national, regional and global experiences of industrial development and urbanization in arid regions in combination with agriculture, cattle breeding and afforestation.

This was stated in two "associated" thematic reports specially prepared for the conference: "*Integrated development of desert and desertification control in the Turkmen SSR*" (A/Con. 74/22) and "*Golodnaya steppe*" (case study of a key site) (A/Con. 78/23). They were included in an official list of conference materials.

23.2 Work with UN Organizations

The Desert Institute is an initiator of international meetings and seminars. Taking into consideration, the great experience of the Desert Institute in researches and training of personnel, in 1978 UNEP took a decision to organize courses for qualification improvement on the issues "Fixation of drifting sands" and "Reclamation of saline irrigated lands". For preparation of a special project it was assigned 40 months (from October 1978 to December 1982). Its implementation started in 1978 within the framework of the project "Land resources of the world". The objectives of this project were:

- Long-term: assistance to developing countries in the arid, semiarid and subhumid zones in desertification control, protection and improvement of the natural environment by way of socio-economic development of regions on the basis of integrated development of industry and agriculture.
- Short-term: assistance to governments of some interested, developing countries in outlining particular development schemes for all or some regions specially identifying types, scale and socio-economic feasibility of proposed measures.

Scientists from the Desert Institute took part in the preparation of most of the study materials on desertification control that were published for the UNEP/USSR Project, "Program of publications and information support". Thus, under the project "Study courses on desertification control" the following study materials were published: "Fixation and afforestation of drifting sands in the USSR deserts", "Problems of salinity control of irrigated lands", "Ecology, management and productivity of pastures", and "Principles and methods of fixation of moving sands". Within the framework of the project, "Desertification control by integrated development", three documents were issued: "*Experience of desertification control in the USSR*", "*Guidelines for preparation of regional schemes of integrated development*", and "*Development of arid territories and desertification control: an integrated approach*".

In September 1982 the Desert Institute held a meeting for the assessment of the UNEP/USSR projects at which they adopted a Memorandum of Mutual Understanding in which it was stated that UNEP thought it necessary to pay special attention to the development of guiding principles and planning concrete measures for appraisal of desertification and its control. The participants believed that projects should be aimed at practical application of accumulated knowledge on the natural environment conservation and the protection of resources in developing countries.

In 1984 the Desert Institute was awarded the silver medal of the UN Environmental Program (UNEP) for successes in studies of arid territories, the development of theory, and practical methods of desertification control and training of personnel from developing countries.

In February 1987 through the decision of ESCATO/UNEP, the Desert Institute was assigned the function as the national center for Researches and Training of Personnel on Desertification Control. The principal goal of the Center is to ensure implementation of the USSR's commitments to participation in international co-operation on desertification control.

In April 1988 in Ashgabad the third meeting of the UN Inter-Agency Working Group on Desertification Control was held on the premises of the Desert Institute. 13 representatives from different subdivisions of the UN system took part in it including UNEP, UNESCO, FAO, WMO, and ECA. At the sessions they discussed problems of training of national desert specialists, discussed approaches of world banks to problems of desertification control and their role in financing measures on long-term development of a territory affected by drought or desertification.

In January 1994 at the regional preparatory conference connected with the International Convention to Combat Desertification in Central Asia, it was decided to assign to the Journal "*Problems of Desert Development*" an international status having introduced in its Editorial Board renowned desert experts from the USA, China, India, Egypt and France.

23.3 International Courses and Seminars

In 1980 a cycle of courses on ecology, management and productivity of pastures was held within the framework of the project "Study courses on desertification control" which replaced the former project "Land resources of the world". All in all 40 people from 18 countries attended these courses. The organizations involved were the Desert Institute, the Botanical Institute of the Kazakh Academy of Sciences, the All-Union Research Institute of Karakul Farming of the USSR Ministry of Agriculture, the Kazakh Research Institute of Karakul Farming, the Kazakh Research Institute of Grassland Agriculture.

In 1987 on the premises of the Desert Institute courses were offered on ecology, management and productivity of pastures, for specialists from member countries of the South Africa Development Coordination Council (SADCC) including Angola, Botwana, Lesoto, Mozambique, Malawi, Swaziland, Tanzania, Zambia, and Zimbabwe. The purpose of these courses was to train 23 specialists in the methods and techniques of rehabilitation of degraded pastures.

Again under the auspices of the Desert Institute in September-October 1988 training courses for representatives of countries of Western Africa were organized. This was done within the framework of the UNEP/USSR International Project "Development of Deserts and Desertification Control". Specialists from 8 world countries participated in these courses.

In October 1988 member-countries of the UN Economic Commission for Africa took part in a seminar and a study tour on the subject "Drought and land desertification". 19 foreign specialists from 16 countries, including 3 representatives from ECA and 1 representative from the Organization of African Unity attended this seminar. The main goal was to get to grips with the experience of arid territories development and desertification control in the USSR, as well as development of a policy in the field of protection of the natural environment of arid territories.

Between 1979-1991 about 600 people from more than 50 countries of Asia, Africa and Latin America improved their qualifications. About 70 foreign lecturers – leading scientists from 14 world countries – took part in the work of these courses. Modules of the courses combined theoretical aspects in the form of lectures and holding seminars, laboratory and field studies. In the course of training the participants were given the possibility to make statements on the subject about the situation in their countries. They visited various scientific and design institutes active in the relevant fields. Field studies were usually in regions most representative in terms of the processes discussed. Feedback from the participants and foreign lecturers prove that such organization of study courses contributed to their success.

On 22-23 September 1994 the National Seminar on Improvement of Water Supply and Sanitary state of Towns and Settlements in Turkmenistan was held. Specialists from the UN Economic and Social Commission for Asia and the Pacific (ESCAP) were invited to it.

Given the importance of steady progress in countries with transitional economies or in newly independent countries on 16-18 August 1995 in Ashgabat the National Seminar on Economic Reform in Turkmenistan was convened with the participation of ESCAP.

On 5-6 March 1996 in Dashkhovuz, a seminar was set up to discuss the state and the prospects of improvement of water supply and sewerage in the Tashauz velajate, in one of the unfavorable regions of the Aral basin. Specialists from ministries and departments, research institutions, higher educational establishments of Turkmenistan, the World Bank, UNDP, UNICEF and others took part in this seminar.

The initiators of the International Seminar "Problems of Steady Development of Pasture Land Use in Central Asia" (4-6 March 1997, Ashgabat, Turkmenistan) were the World Bank (the Institute of Economic Development), the Inter-State Council on the Aral Sea, the Government of the Netherlands and the Desert Institute. Scientists and experts on pastures from Kazakhstan, Uzbekistan, Kyrgyzstan, Tadjikistan and Turkmenistan took part in the work of the seminar.

Scientists as representatives of the World Bank were invited to the seminar. Among them were Jose Fourtado, the leading expert on the natural environment of the Institute of Economic Developnet of the World Bank; Shilhorn Chat, from the World Bank's Department of Agricultural and Natural Resources; Wintford Peter, the Group for Promotion of Saving the Aral Program; Dankova Rimma Nikolayevna, a representative of the Institute of Economic Development of the World Bank; and Nazari Dashlibron Purinazar from the Islamic Republic of Iran..

At the seminar one representative from each Central Asian state made a report, after which the other seminar participants voiced their comments and personal opinions on all discussed questions. Rapporteurs and speakers paid special attention to the role of private farmers, cattle breeders and farmers' associations in ecologically constant pasture land use, the participation in various forms of property and landholding, the density and management of pastures, the role of local authorities (central, velajate and etrape) in the use of pasture lands and lastly, particular problems requiring the development of new approaches to resolving these burning problems.

Director of the Desert Institute Professor Babayev who made a review report on the rational use and steady development of pasturelands and desertification control problems took an active part in the work of the seminar.

Conclusions of the seminar pointed to the necessity to adopt in Turkmenistan a law on pastures. It is necessary to outline the scale of surface or deeper land improvement and the water supply of pastures, to study redistribution of pastures, to establish seed growing farms, and to obtain state support for the projects, and many other problems that in the near future will be put into practice.

In connection with ratification by the *Medjliss* (Parliament) of Turkmenistan of the UN Convention on Desertification Control in the last two years, the Desert Institute prepared a number of project proposals for various international organizations to implement projects on desertification control with the participation of organizations from developed countries that signed the Convention.

23.4 Development Assistance in the South

In 1983 Desert Institute scientists together with specialists from the Research Institute of Gosgrazhdanstroy took part in the preparation of a plan of integrated, socio-economic development of one of the arid regions in Peru. The scheme was an analytical overview of the state of the natural resource based economy and social conditions in the Ika Department, and the nature and intensity of desertification processes on its territory. Using a mathematical model this blueprint proposed several options of economic development in the department taking into consideration ecological and social factors, as well as the integration of Ika into the process of general economic development of the country. The scheme was prepared on the basis of data obtained in field investigations in Peru and interpretation of statistical, cartographic and other materials presented by the Peruvian team. The scheme was submitted to the Government of Peru in August 1986.

In 1984 they prepared a regional scheme for integrated development of a region in Timbuktu, the Republic of Mali. The scheme was prepared jointly by the Desert Institute and V/O "*Soyuzvodproject*" of the USSR Ministry of Land Reclamation and Water Management. The scheme contained proposals on the development of land irrigation, the water supply of pastures, water supply of settlements, desertification control in the territory of eight administration regions in the north of Mali with a total area of 5,000 km^2. In 1987 the scheme was submitted to the Government of Mali.

23.5 Cooperation with Iran

In November 1995 the Desert Institute and the Department of Forest and Pasture Studies at the corresponding Iranian Minisitry (*Djakhad-e-Sazandegi*) signed the Agreement on Scientific and Technical Cooperation that envisaged organization of joint R & D on the problem "The Deserts of Iran and Turkmenistan and Desertification Control Strategy".

The Turkmen side has undertaken the preparation of programs and methodological materials on:

- the ecology and phytoreclamation of desert pastures;
- the fixation and aforestation of drifting sands;
- the investigation of local waters and their use;
- the genesis and classification of soils in Iran;
- the mapping of processes of anthropogenic desertification;
- the creation of an information data bank on problems of deserts.

23.6 Cooperation with Israel

An Agreement on Cooperation was signed between the Desert Institute and the Desert Institute of the Ben-Gurion University (Israel). This takes into consideration the UN Convention on Desertification Control and significant experience accumulated in Turkmenistan and Israel on the control of such negative phenomena as wind and water erosion, soil salinity, diminished productivity of pastures. Out of this joint work, the development of agricultural practices is envisaged. For bilateral cooperation the following problems are of greatest interest:

- the ecological foundations of enhancing productivity of desert pastures;
- the control of wind and water soil erosion;
- the application of progressive technologies in the irrigation of agricultural crops and selection of new varieties for desert development;
- the water supply of desert territories;
- afforestation of desert lands;
- the harnessing of solar and wind energy for power supply of small users;
- the development of cartographic criteria and preparation of desertification maps;
- the training of personnel in desertification problems.

It was also planned to undertake preparation and implementation of joint international projects connected with training of specialists from developing countries on the territory of Turkmenistan and Israel with participation in researches on the territory of other countries affected by desertification processes.

23.7 Cooperation with China

For many years the Desert Institute has cooperated with scientific organizations of the Chinese People's Republic. The main partners in this cooperation on the Chinese side are the Desert Institute of the Chinese Academy of Sciences (Landzhou) under 1993 Agreement on Joint Activities; with the Institute of Soil and Desert Studies (Urumchi) the Agreement on Friendly Cooperation of 1994; and the Forestry Institute of An Suar (Urumchi) under the 1992 Protocol on Cooperation.

In 1993 the Desert Institute together with the Desert Institute of the Chinese Academy of Sciences established the Scientific Center "Barkhan" within the framework of which appropriate works are carried out.

23.8 Support from Germany

The Ministry of Economic Cooperation of Germany together with the Government of Turkmenistan outlined a concept on nature conservation and rational use of natural resources as a possible alternative sphere of cooperation. One of the lines of such cooperation is support by the German Agency on Technical Cooperation (GTZ) for the activities of the Desert Institute within the framework of the International Convention to Combat Desertification.

In order to familiarize the world community with scientific projects implemented by the Desert Institute over the course of more than 30 years, a collection of abstracts was prepared which made various proposals: for the activation of initiatives from below; on conducting seminars in rural areas and other measures to educate the population of arid zones in desertification control.

To publicize the results of the research activities of the Institute, a collection of abstracts of scientific works was prepared. This collection is intended for wide public dissemination in the arid zones: this strategy is consistent with the recommendations of the UN Convention on Informing the World Public on Implemented Scientific Projects. The intention is to attract and activate grassroots initiative from below to further desertification control. Presenting the material in a popular form this collection gives a possibility to organize seminars in rural areas with a view to enhance ecological knowledge of the population and mobilization of location natural resources for solution of desertification problems at a local level.

International cooperation undertaken by the Desert Institute is fully in accord with provisions of Clause 12 of the UN Convention to Combat Desertification which provides for the necessity to create favorable conditions both at the regional and subregional levels. This cooperation should cover such spheres as transfer of both new and traditional technologies, and the widespread provision of information of the results of researches and advanced practical developments.

References

Babaev AG and Freikin ZG: Deserts of the USSR, yesterday, today, tomorrow. Mysl Publ. Moscow.

24. The Consulting Experience of the Desert Institute

C.O. Muradov, A.A. Khachaturyan

ABSTRACT: The article provides a convenient overwiew of the Desert Instute's work and configuration. It was officially founded in 1962, and quickly emerged as an important center in this field. The main lines of research there are briefly listed, along with the organizational structure. The functions of the ten associated laboratories are mentioned, as well as associated research and coordination centers. The projects of the Institute are listed.

The Desert Institute has accumulated unique experiences. The real date of the origin and development of desert science in Turkmenistan – which the Desert Institute now has the honor to represent – should be considered 1912, the year when the Repetek sandy desert station was set up and where the first path-breaking studies of sandy deserts were started.

The Desert Institute officially began to function in 1962. It was established in order to concentrate the activities of scientists, and as a means of development of scientific and practical foundations for the integrated study and rational use of natural resources in the arid territories.

Already within five years it had acquired the status of the Head Research Center on the problem "Integrated study and development of deserts in Central Asia and Kazakhstan"; later this framework was still further widened to cover the whole of the former Soviet Union. The main lines of researches conducted at the Institute were:

- the development of remote sensing methods of study, and monitoring of the natural conditions in deserts;
- the mapping of desertification processes;
- the investigation of atmospheric events affecting the life of people and the production sphere;
- the study of the regularities governing formation and development of desert relief;
- the development of methods to control sand drifts and deflation of key buildings and installations;
- the development of phytoreclamation and engineering methods of stabilization, afforestation of sands and growing plants on them;

- the development of methods to enhance and predict productivity of pastures;
- the classification, diagnostics, mapping and assessment of desert soils;
- the development of methods of rational management of local water resources;
- the development of methods of drainage water use;
- the development of desalinization and water treatment complexes;
- informational support for research and the creation of databank;
- the training of highly skilled personnel in the desert science and desertification control;
- international cooperation and implementation of the UN Convention on Desertification Control.

A flexible system of management of the scientific potential makes it possible – depending on research objectives – to alter operatively the structure of the Institute where at present more than 160 people work, among them more than 60 scientists with candidate and doctoral degrees.

24.1 The Desert Institute's Laboratories

Today the Institute unites 10 laboratories (Fig. 106) making up its scientific basis. The Institute includes the Repetek International Biosphere Nature Reserve and the sandy desert station, the Nebitdag Afforestation Station and 6 scientific stations located in different regions of Turkmenistan. Brief information about them is given below.

24.1.1 The Laboratory for Monitoring of Desertification Processes

Application of remote sensing in studies of arid lands and monitoring of desertification processes; study of optical properties of arid geosystems and other factors that are responsible for specific features on photo images of landscapes on aerospace photographs; development of thematic interpretation of aerospace photographs, and preparation of thematic maps.

24.1.2 The Laboratory on Drifting Sands

Study of the regularities governing formation and development of drifting sands and physical foundations of deflation; development of methods to control sand drifts and deflation on major engineering structures; development of theory and practical foundation of fixation of moving sands.

24.1.3
The Laboratory of Water Resources

Investigation of local water resources, development of rational methods of their management, protection and renewal; qualitative and quantitative assessment of local waters and development of methods for their collection, storage and efficient use; development of small oasis farming, and planting trees and shrubs using local surface runoff.

24.1.4
The Laboratory of Pasture Improvement

Establishing ecological principles of improvement of a forage base of deserts and enhancement of productivity of deteriorated pastures; investigation of ecological and biological specific features of growth and development of pasture phytocenoses of the arid lands on the basis of long-term overground monitoring; and the development of methods of radical and surface improvement of different types of pastures.

24.1.5
The Laboratory of Pasture Mapping

Mapping of natural pastures and preparation of maps for different purposes applying geoinformation systems (GIS); quantitative and qualitative assessment of a nutritive value of pasture vegetation.

24.1.6
The Laboratory of Pasture Agroforecasting

Short-term and long-term forecasting of pasture productivity; development of methods for assessment and long-term forecast of pasture productivity with regard to agrometeorological factors; development of differentiated models of long-term agrometeorological forecasts of pasture yields.

24.1.7
The Laboratory of Soil Biogeochemistry

Genesis and classification of flat and mountain soils; geochemical and biochemical analysis of the soil composition; appraisal of the natural, geochemical background of the pedosphere of deserts and fundamental description of its composition; development of methods for appraisal of soil fertility and productivity and the role of soils in the biosphere of deserts.

24.1.8 The Laboratory of Sand Phytoreclamation

Phytoreclamation of sands and protective afforestation, plant growth on sands and sandy lands using progressive technologies of irrigation; introduction of high-productive forage and fruit-bearing crops; development of phytoreclamation methods for control drifting sands nearby the most important buildings and installations.

24.1.9 The Laboratory of Water Desalinization and Treatment

Studies of local sorbents for treatment and desalinization of water in the systems of individual and collective water use; discovering natural sorbents and materials from which they can be obtained; and the development of water treatment units.

24.1.10 The Laboratory of Scientific, Technical Information

Informational support of research works and the creation of databank on arid lands; creation of a geoinformation system; inventions and proposals for improving production methods; and a patent search on themes of the most important researches.

24.1.11 Department of International Scientific Relations

Participation of the Institute in implementation of international scientific programs on desertification control; organization and holding of international training courses, seminars, conferences, symposia; promotion of scientific relations with foreign partners; preparation of design proposals and programs of scientific cooperation.

24.2 The Desert Institute's Research Stations

24.2.1 Repetek International Biosphere Nature Reserve and Sandy Desert Station

Investigation of regularities governing long-term dynamics of landscapes in sandy deserts and specifics of their formation; integrated study of structural-functional organization of sandy desert biogeocenoses for development of the methods to control a life of communities; revealing of potential possibilities for their use.

24.2.2 Nebitdag Afforestation Station

Investigation of bioecology and introduction of valuable forest, fruit-bearing and forage crops into major phytocenoses, as well as study of afforestation conditions on drifting sands, takyrs and takyr-like soils for growing on them of forest and forage crops.

24.2.3 Regional Center of Research and Training of Personnel on Problems of Desertification Control

This centre functions within the framework of ESCAP/UNEP and studies international experience on research and development of deserts; practical implementation of international scientific projects on rational management of resources of arid lands and control of desertification processes.

24.2.4 Scientific Information Center "Aral"

Information Support for the Commission on steady development of the Inter-State Council on the Aral Sea Problems, as well as national and sub-regional projects within the framework of the UN International Convention on Desertification Control.

24.3
International Programs and Projects

For more than thirty years the leading scientists and experts of the Institute have actively participated in implementation of international projects of UNEP, UNESCO, ESCAP, FAO and other UN subdivisions related to problems of study and development of arid lands, as well as the problem of desertification control.

The international scientific relations of the Institute are very wide and variegated. Since 1967 the Regional Scientific Council has been functioning at the Institute. It includes leading scientists from Kazakhstan, Kyrgyzstan, Tadjikistan, Turkmenistan, Uzbekistan, Azerbaijan and the Russian Federation. The Council coordinates research efforts on arid problems of many research institutions and higher educational institutions in the Central Asian countries.

For a long time, the Institute has actively cooperated with scientists and experts from Iran, Syria, Pakistan, India, China, Mongolia, USA, Israel and other countries in coping with problems related to the study and development of resources of arid lands. The Institute regularly organizes and holds international, scientific conferences, symposia and seminars which discuss strategic problems of integrated research and rational management of resources of arid territories, as well as control of desertification processes.

The Institute was awarded the Order of the Labor Red Banner in 1969 for its considerable contribution to the development of theory and practical recommendations for the development of arid lands, and in 1984 it was awarded the Silver Medal of the United Nations Environmental Program (UNEP).

At present engineers, representatives of industry and of different educational or research centers in Turkmenistan and foreign countries apply to the Desert Institute for scientific and/or practical consultations on desertification control.

24.4
The Desert Institute's Most Important Findings

Among the most significant theoretical and practical developments, one should mention the recommendations of the Institute on protection from sand drifts and deflation near important engineering structures.

Of great practical value are the results of scientific researches over many years by the Desert Institute which were widely applied in the design and construction of the Karakum canal. A scientific forecast prepared by the Institute of the likely ecological impact in connection with putting the canal into operation allowed the development of valuable scientific/practical recommendations on the development of virgin lands.

In 1981 a group of scientists of the Desert Institute was awarded the USSR State Prize for implementing the technology of enrichment of desert pastures in the practice of Karakul farming in Central Asia. Maps of pastures in Turkmenistan and Central Asia made as a result of the Institute's researches are widely used in planning cattle population and the organization of its grazing.

Analysis of the results of lengthy researches and the in-depth study of traditional methods of water supply to the Karakum led employees of the Institute to suggest a number of original proposals on methods of water supply to desert pastures. The most interesting is the development of a technology of artificial formation and storage of freshwater lenses as a result of collecting runoff from takyrs. As well as that were many different designs: for helio-windmill, power plants combined with desalinization units; water lift units intended for pasture water supply; and a whole wealth of unique units for the treatment and desalinization of fresh water using local sorbent materials.

One can designate the development of recommendations on planting trees and shrubs in settlements and the creation around them of forest belts as a considerable contribution of the Institute. Widely applied too are developments of the Institute on plant growing on sands and takyrs using up-to-date methods of irrigation and the introduction of highly productive crop varieties.

The Institute together with some interested organizations and with the financial support of UNEP developed guidelines on the preparation of regional schemes on integrated development of problems of desertification control. The schemes generalize and systematize major principles of an integrated approach to planning of the socio-economic development of territories in order to control desertification and to protect arid ecosystems.

Scientists of the Institute were invited as experts and directly participated in preparation of the schemes of integrated development for different countries. Implementation of such schemes makes it possible to consider the areas they cover as model regions on the basis of which the proposed sets of measures on development of analogous territories can be tested.

An important place in consulting activities of the Desert Institute is given to assistance to interested countries in the arrangement of scientific research and training of scientific personnel on problems of desertification control. An apposite example here is long-term cooperation with scientists in China, Mongolia, Syria, Iran and other countries. Some consulting works done under such projects are:

1994-1998	Desertification control in Iran
1991-1995	Comparative, ecological characteristics of desert territories in India and Turkmenistan and desertification control
1989-1991	Amudarya-Colorado: past, present, future (USA-Turkmenistan)
1987-1990	Support of West African countries in their activity to control desertification. Case study of Mali
1992-1994	Study of desertification processes and protection of arid ecosystems in Mongolia
1990-1991	Revival of saline soils in the Euphrates river basin (Syria)
1983-1986	Desertification control through integrated development (Peru)
1979-1980	Soil ecological investigations of the pasture zone in the Libyan Arab Jamahiriya

1979-1980	Investigation and mapping of soil and pasture resources in coastal areas (Algeria)
1967-1968	Control of drifting sands in the construction zone of Bagdad-Basra (Iraq)

Consultations concerning scientific methodology and practical recommendations by the Institute made it possible in China to establish within a short time period a powerful base for the staging and implementation of multipurpose researches into aridity problems, and enabled Chinese scientists to reach great heights in desert science.

Results of joint efforts by the Institute's staff with their Mongolian colleagues can be considered a weighty contribution to the development of Mongolian desert science. Extensive scientific materials obtained in the course of joint field expeditions allowed the preparation of recommendations on afforestation and the fixation of drifting sands in Mongolia, as well as on control of sand drifts and protection of engineering structures from deflation. Apart from this, with the active participation of the Institute a map was prepared with risks of desertification for arid territories of Mongolia, and recommendations developed on the protection of arid ecosystems and rational management of their natural resources.

Expert groups made up of leading scientists based at the Institute have worked at different times in Algeria and Libya. With their direct participation a methodology was developed for the preparation of an ecological map of Central Libya based on aerospace photographs and thematic maps. This methodology makes it possible to assess a complex of factors determining productivity, stability and anthropogenic disturbance of ecosystems. The results of this work can also be used in preparation of ecological maps of other arid regions.

The Desert Institute provided scientific-methodological assistance to the Afghanistan Center of Natural Sciences in studies on the introduction and adaptation of the most valuable, agricultural crops.

The Desert Institute provides consulting services in training and improvement of qualification. In the period from 1978 to 1992 in the course of implementation of the UNEP projects on control of desertification processes more than 600 specialists from developing countries of Asia, Africa and Latin American managed to improve their skills at the Institute. In recent years representatives from Syria, Egypt, Iran, Mali, China, Mongolia and other countries attended the Institute as interns or post-graduate students.

Experience of the Institute was used in the course of preparation of the Regional and National Programs of Action on Desertification Control.

25. Conclusion

A.G. Babaev

At present combating desertification as a serious ecological problem has a global importance. Researches have showed that deserts are very vulnerable under anthropogenic influence, and that is why any unreasonable steps may result in the development of desertification.

But rapid reaction and adoption of urgent, practical scientific steps allow the neutralization or prevention of first indications of desertification especially in sandy deserts, in which the dynamic ecosystems balance is less sustainable and the processes of self-regulation are more difficult.

Desertification is a combination of natural and anthropogenic processes which in total have resulted in the destruction of all links of ecological system and sharp reduction of the Earth's natural resource potential.

Desertification is a complex, many-sided process having both geographical and social-economic aspects. Desertification's degree and scales depend upon the intensity of natural processes and anthropogenic factors.

Socio-economic factors are very important: the possibility of state planning of the economy and regulation of the influence on a desert's ecosystems are of great significance to prevent desertification; the economic structures and the degree of their management. The real possibilities and material means, that are necessary for achieving measures on protection, restoration and increasing productivity of natural landscapes, depend on the state system.

From our point of view, the UN Convention to Combat Desertification adopted in 1994, is regarded as the most global project of the century because of its scientific scope and power.

It is important to stress that the UN through its international organizations can attract serious public attention to the processes of desertification. At present millions of people, first of all, inhabitants in arid areas follow with emotion and hope the development of international cooperation in the practical realization of global, regional and national projects to stop further development of desertification processes.

Admitting the threat of the development of desertification processes to be real and attaching great importance to the development of methods of rational usage of natural resources of arid lands and protection of their nature, in 1967 in the former Soviet Union the "Scientific council on the complex study and deserts' development" was created on the basis of the Desert Institute with the center in Ashgabat (Turkmenistan).

The Scientific council has achieved important results in many fields: the elucidation of melioration of interactions of natural and anthropogenic factors of desertification; the development of methods of long-term prognoses of natural processes causing desertification risk; the field of theory of managing and regulation of phenomena in the environment on various levels. Its conception and strategy of combating desertification were developed and adopted by the scientific public.

In the near future, anthropogenic influence upon deserts will increase. Therefore to prevent desertification's processes, it is necessary to plan measures on protection of desert's nature and demand restoration of lost components of natural habitats.

Through mapping of present and possible desertification, investigations might reveal areas which have undergone desertification, define its scale and degree of spreading, causes and consequences. With a view to early discovery of desertification processes and timely measures to prevent them, a set of available indicators was demonstrated by us.

Taking into account that desertification is an inter-disciplinary problem, the development of indicators has required the coordination of efforts of various experts in the field of social, agricultural geographical and physical-mathematical sciences that allows us to foresee the course of changes of natural ecosystems and take steps to prevent the beginning of natural processes which are negative for the economy.

Further studies must be directed to the development of complex measures on rational usage of natural resources and restoration of degraded landscapes of arid lands. Harmonic combination of measures on rational development of natural resources in crucial zones for desertification and protection of them from further destruction may stop and save them from the desertification processes.

The notion of desertification is broad enough. It includes some aspects which may be divided into two groups:

- Natural desertification processes, marked by progressive drying up of arid territories, including climatic, hydrogeological, morphodynamic, soil, plyto and zoogene processes.
- Anthropogenic desertification processes as a result of intensive and irrational agriculture, salinization, waterlogging, livestock overgrazing of pastures, annihilation of forests, urbanization, creation of engineering construction, population migration etc.

The UN Convention envisages the development of an integrated complex of measures to prevent desertification's processes. But it requires large investments and availability of trained personnel. Therefore it is necessary to define primary national and regional tasks for the implementation of the UN Convention to Combat Desertification. Furthermore, types of desertification's processes, their intensity, scale and rate are different in various regions. It also requires a differentiated approach to the problems of combating desertification in the countries of Central Asia having regard to specific natural and socio-economic conditions.

In the UN Convention it is correctly underlined that desertification is not a problem that can be solved fast. It requires constant observations and scientific analysis, planning and long-term prognosis. Long-term prognosis of natural processes in places of intense anthropogenic influence on natural environment must be of practical significance. It would be the result of the application of complex, physical geographical researches with the usage of mathematical and physical methods including modeling of natural processes.

In connection with the UN Convention to combat desertification and the UN Convention on biodiversity, ratified by Medjlis (Parliament) of Turkmenistan on June 18, 1996, in future the Institute's work will be forming a strategy of nature use and environmental protection. For all this the main tasks of the Institute will be:

- carrying out of fundamental and applied scientific investigations in the field of environmental protection and the rational use of natural resources;
- realization of scientific projects, development of recommendations and standards for the restoration of forest and grazing lands, flora and fauna protection, combating desertification and environmental monitoring;
- introduction of a more scientific dimension to regional and world achievements in economy;
- strenghtening of cooperation with ministries, departments and organizations on the implementation of complex, scientific programmes;
- involvement of foreign investments for the implementation of great scientific projects.

For the implementation of these tasks, the ecological scientific information center was established. Here there is a well-constructed system of collection, storage and treatment of the information about the environmental and natural resource potential not only of Turkmenistan but also of other countries of Central Asia. For all this the newest GTZ technology and global networks will be used. Already the Desert Institute has homepages in the Internet system.

At the present stage, the Institute has acquired a profound understanding of natural conditions, the resources of arid lands to combat desertification, the training of personnel on arid problems, and the introduction of scientific recommendations on the wider use of the experience of the wise. It will allow us in future to give skilled consultations to ministries and departments of the country of various projects and plans, to speak as experts on elaboration of operative solutions, and to be coordinators of international programs to combat desertification and further the conservation of biodiversity.

Effective solutions to combat desertification's problems cannot be realized outside of the context of major socio-economic and scientific challenges, and the wide usage of progressive technology with compulsory restriction of undesirable consequences for the environment.

The maintaining of sustainable rates of economical and social development of any country is associated to a considerable extent with the preservation of ecological well-being. The economic mechanism with full regard of ecological factors will to a great extent determines a state's sustainability.

Annex

Fundamental Scientific Works of the Desert Institute of the Academy of Science Turkmenistan

Unless stated otherwise, the publications are in Russian

1963

Babaev AG: Karakum Desert. In: Fedorovich BA (ed.). Ashkhabad, AS TSSR Publ.
Rabochev IS (ed.): The desert land-water resources and their utilization.
Levadnyuk AT: Sand tracts in northern Western Turkmenistan lowland. In: Sidorenko AV.
Nechaeva NT and Prikhod'ko SY: The biology and crops of feruls and dorems in Turkmenistan.
Novikov GS (ed.): Experience of work of Repetek sandy desert station.
Nechaeva NT, Babaev AG (ed.): Natural conditions, stock breeding and fodder supply of deserts. Collected articles.
Kolodin MV and Smirnov LN (ed.): Minerals, power resources of deserts and their utilization. Pulication of AS TSSR, Ylym, Ashkabad.

1964

Babaev AG (ed.): Instructions for plant-growing development on takyrs and takyric soils on the base of local, surface water flow. Publication of Arid lands. AS TSSR Publ., Ylym, Ashkhabad.
Pietrov MP: The USSR deserts and their development. Nauka Publ, Moscow, Leningrad.
Rabochev IS: Melioration of saline soils in Amudarya middle flow. AS TSSR Publ., Ashgabat.
Efimov GS: Water balance of the Tedjen oasis. In: Rabochev IS (ed.) AS TSSR Publ., Ylym, Ashkabad.

Nechaeva NT and Prikhod'ko SY: Artificial winter pastures in the piedmont deserts of Central Asia. In: Shalyt MS (ed.). Turkmenistan Publ., Ashkhabad.
Pietrov MP: The deserts of Central Asia vol.1: The Ozdos Alashan and Beishan. In: Sinotsin VM (ed.). Nauka, Moscow, Leningrad.

1967

Rabochev IS (ed.) Genesis, geochemistry and melioration of TSSR saline Soils. AS TSSR Publ., Ylym, Ashgabat.
Efimov GS Water balance and zoning, for drainage in TSSR. AS TSSR Publ., Ylym, Ashgabat.
Kolodin MV *et. al.*: Modern methods of water desalinization. In: Klyachko VA (ed.) AS TSSR Publ., Ylym, Ashgabat.
Pietrov MP (ed.): The complex study and development of Turkmenistan deserts. AS TSSR Publ., Ylym, Ashgabat.
Mukhamedov G: Alive desert. Turkmenistan Publ., Ashgabat.
Mashrykov KK (ed.): Irrigation agriculture and desert development. Collected articles. AS TSSR Publ., Ylym, Ashgabat.
Rabochev IS and Grinberg LM and Lavrinenko VT: Development of irrigation agriculture in Turkmenistan. AS TSSR Publ., Ylym, Ashgabat.

1968

Babaeva AK (ed.): Adaptation of man and animals to desert conditions. Theses of reports of All-Union scientific conference on study and development of Central Asia arid lands in Kazakhstan. AS TSSR Publ., Ylym, Ashgabat.
Berdyklychev K: Meliorative importance of arboreous species in Northern Turkmenistan. In: Al'berskyi AV (ed.) AS TSSR Publ., Ylym, Ashgabat.
Tashliev AO (ed.): Desert biocomplexes and increase of their productivity. "Animal World"/ Theses of reports of All-Union scientific conference on study and development Central Asia and Kazakhstan. AS TSSR Publ., Ylym, Ashgabat.
Nepesova NT (ed.): Desert biocomplexes and increase of their productivity. "Vegetation"/ Theses of reports of All-Union scientific conference on study and development Central Asia and Kazakhstan. AS TSSR Publ., Ylym, Ashgabat.
Kalugit PJ (ed.): Water resources. Theses of reports of All-Union scientific conference on study and development Central Asia and Kazakhstan. AS TSSR Publ., Ylym, Ashgabat.
Nepesova NT and Manakov VS (ed.): Fodder base of deserts and intensification of cattle-breeding. Theses of reports of All-Union scientific conference on study and Kazakhstan. AS TSSR Publ., Ylym, Ashgabat.
Rabochev IS (ed.): Melioration of Turkmenistan desert lands. AS TSSR Publ., Ylym, Ashgabat.
Pietrov MP and Nepesova NT (ed.): Shifting sands and land forest amelioration. Theses of reports of All-Union scientific conference on study and Kazakhstan. AS TSSR Publ., Ylym, Ashgabat.
Rabochev IS (ed.): Soils and land funds. Theses of reports of All-Union scientific conference on study and Kazakhstan. AS TSSR Publ., Ylym, Ashgabat.

Babaev AG: Natural conditions and zoning of desert territories. Theses of reports of All-Union scientific conference on study and Kazakhstan. AS TSSR Publ., Ylym, Ashgabat.
Baum VA (ed.): Power resources of desert. Theses of reports of All-Union scientific conference on study and Kazakhstan. AS TSSR Publ., Ylym, Ashgabat.

1969

Aranbaev MP: Serozems and light-brown drysteppe soils of Central Kopetdag. In: Kovda VA (ed.). AS TSSR Publ., Ylym, Ashgabat.
Babaev AG, Orazov A, Muradov EM and Freikin ZG: Turkmenistan. In: Ed. Kunin VN (ed.). Mysl Publ. Moscow
Pietrov MP and Babaev AG (ed.): The complex study and development of Central Asia and Kazakhstan desert territories. AS TSSR Publ., Ylym, Ashgabat.
Leshchinskyi GT: Rains and cloud-bursts on TSSR territory. AS TSSR Publ., Ylym, Ashgabat.
Redjepbaev K: Tedjen delta's soils and their agricultural utilization. AS TSSR Publ., Ylym, Ashgabat.

1970

Kirsta BT: Flow of suspended alluvions of Turkmenistan rivers. In: Shults VL (ed.). AS TSSR Publ., Ylym, Ashgabat.
Kopekov CB: Northern Turkmenistan landscapes and their economic assessment. AS TSSR Publ., Ylym, Ashgabat.
Leshchinskyi GT: Rational utilization of temporary surface water flow for irrigation of desert pastures. In: Babaev AG (ed.). AS TSSR Publ., Ylym, Ashgabat.
Nepesova, NT (ed.): Central Karakum vegetation and its productivity. AS TSSR Publ., Ylym, Ashgabat.

1971

Sultanov FF (ed.): Adaptation of man and animals to desert conditions. AS TSSR Publ., Ylym, Ashgabat.
Nepesova NT and Rustamov AK (ed.): Desert biocomplexes and increase of their productivity. AS TSSR Publ., Ylym, Ashgabat.
Bobkov AG: Territory of Turkmenistan in conceptions of ancient scientists. Turkmen Statistics Department. Ashgabat.
Rabochev IS and Kalugin PI (ed.): Land-water resources of deserts. AS TSSR Publ., Ylym, Ashgabat.
Babev AG (ed.): Complex investigating, removing and development of desert of Turkmenistan. AS TSSR Publ., Ylym, Ashgabat.
Orlovskyi NS: Evaporation from water surface of small reservoirs of Turkmenistan. AS TSSR Publ., Ylym, Ashgabat.
Babaev AG (ed.): Natural conditions and shifting sands of deserts. AS TSSR Publ., Ylym, Ashgabat.

Sanin SA: Agrochemical and chemical properties of takyr and takyric soil of Murgab oasis. In: Paletskaya LN (ed.). AS TSSR Publ., Ylym, Ashgabat.

1972

Aranbaev MP, Malaev NR and Niyazova MM: Mineralogical composition of serozems and brown soils of southern Turan facies. In: Kovda VA (ed.) AS TSSR Publ., Ylym, Ashgabat.

Batyrov A: Rural settling at Lenin Karakum canal zone. In: Babaev AG and Freikin LG (ed.). AS TSSR Publ., Ylym, Ashgabat.

Ivanov AP: Physical principles of sands deflation of desert. In: Dobrin LG and Orlovskyi NS (ed.). AS TSSR Publ., Ylym, Ashgabat.

Mukhamedov G: *Ephedra strobilaceae* in Karakum deserts. In: Nardina, NS (ed.). AS TSSR Publ., Ylym, Ashgabat.

Nikolaev VV: Natural fodder resources of Turkmenistan. Nechaeva NT (ed.). AS TSSR Publ., Ylym, Ashgabat.

Babaev AG (ed.): Experience of study and development of Eastern Karakum. AS TSSR Publ., Ylym, Ashgabat.

1973

Ataev AA, Batyrov AB and Freikin LG: Main forms of economic efficiency increase of agricultural desert development of Central Asia. In: Nechaeva NT (ed.). AS TSSR Publ., Ylym, Ashgabat.

Babaev AG: Oasis sands of Turkmenistan and methods of their development. In: Pietrov MP (ed.). AS TSSR Publ., Ylym, Ashgabat.

Kalenov GS: Vegetation of low-lying Karakum in connection with soil and subsoil conditions. In: Nechaeva NT (ed.). AS TSSR Publ., Ylym, Ashgabat.

Kolodin MV and Seyitkurbanov S: Desalinization of saline waters. In: Khaliev NP (ed.). AS TSSR Publ., Ylym, Ashgabat.

Lavrov AP: Essay of Zaunguz Karakum soils. In: Babaev AG (ed.): AS TSSR Publ., Ylym, Ashgabat.

Nechaeva NT, Vasilevskaya VK and Antonova KG: Living forms of Karakum desert vegetation. Nauka, Moscow.

Babaev AG (ed.): Principal problems of complex investigations of Central Asia and Kazakhstan desert territories. AS TSSR Publ., Ylym, Ashgabat.

Pietrov MP: The Earsh's Deserts. Nauka, Leningrad.

Rabochev IS, Ovsyannikov AS *et. al.*: Desalting of soils and irrigation of agricultural crops with mineralized waters. In: Rabochev IS (ed.). AS TSSR Publ., Ylym, Ashgabat.

Cherednichenko VP: Morphology of eolian relief and construction of pipelines in desert. In: Pietrov MP (ed.). AS TSSR Publ., Ylym, Ashgabat.

1974

Orlovskyi NS and Volosyuk LI (ed.): Agroclimatic resources of TSSR. Gidrometeoizdat, Leningrad.

Aranbaev MP, Rabochev IS *et. al.*: Primary productivity, the dynamics and organic matter's balance of Murgab oasis agricultural crops. In: Kovda VA (ed.). AS TSSR Publ., Ylym, Ashgabat.

Ataev A: Economics of irrigation of South-Eastern Karakum pastures. AS TSSR Publ., Ylym, Ashgabat.

Nechaeva NT and Nikolaev VN The map of Turkmenistan pastures. Ashgabat.

Babaev AG (ed.): The complex investigating, remaking and development of Turkmenistan deserts. AS TSSR Publ., Ylym, Ashgabat.

Lavrov AP, Larin EV *et. al.*: Soil - climatic zoning of Karakum Canal area. In: Rabochev IS (ed.). AS TSSR Publ., Ylym, Ashgabat.

Lalymenko II: Sprouting seeds of arboreous and shrub vegetation under different salinization. AS TSSR Publ., Ylym, Ashgabat.

Leshchinskiy GT: Resources of temporary surface water flow of Central Asia and Kazakhstan deserts. In: Kunin VN and Babaev AG (ed.). AS TSSR Publ., Ylym, Ashgabat.

Orlovskyi NS and Volosyik LI: Weather and distant pasture Eattee breeding of Turkmenistan. In: Babaev AG (ed.) AS TSSR Publ., Ylym, Ashgabat.

Sidorenko DI: Economic efficiency of pastures improvement in Turkmenistan piedmont deserts. Necheva NT (ed.). AS TSSR Publ., Ylym, Ashgabat.

Kharin NG: Anglo-Russian glossary of geographical terms on deserts. In: Pietrov MP (ed.). Ashgabat.

1975

Artykov K: Agroclimatic bases of phytomelioration works in Turkmenistan foothills. In: Ovezliev AO (ed.). AS TSSR Publ., Ylym, Ashgabat.

Berkutenko AN: Determination of vegetation of Repetek reserve. AS TSSR Publ., Ylym, Ashgabat.

Nechaeva NT (ed.): Biogeocenosises of Eastern Karakum. Characteristics of main components. AS TSSR Publ., Ylym, Ashgabat.

Kiril'tseva AA: Phytophenological mapping with application of biometrical methods. In: Pietrov MP (ed.). AS TSSR Publ., Ylym, Ashgabat.

Kirsta BT: Mineralization of water chemical flow of Turkmenistan rivers and methods of their calculation. In: Babaev AG (ed.). AS TSSR Publ., Ylym, Ashgabat.

Kolodin MV: Problems of water desalinization on the Earth. Znanie Publ., Moscow.

Rabochev IS (ed.): Methods of study of mineralogical composition and organic substance of soils. AS TSSR Publ., Ylym, Ashgabat.

Orlovskyi NS: Agroclimatic conditions of cotton growing at Lenin Karakum canal zone. In: Babushkin LN (ed.) AS TSSR Publ., Ylym, Ashgabat.

Kalenov GA: Recommendations for using of vegetative indicators at geological surveys in Karakum. Ashgabat.

Sidorenko DI: Peculiarities of capital expenditure's calculation on TSSR pasture irrigation. AS TSSR Publ., Ylym, Ashgabat.

Kharin NG: Remote sensing for vegetation study. Nauka Publ. Moscow.
Kharin NG and Pietrov MP: Glossary of terms on natural condition and development of deserts. In: Babaev AG (ed.). AS TSSR Publ., Ylym, Ashgabat.

1976

Babaev AG: Short survey of Turkmenistan nature (in Russ. and Engl.). AS TSSR Publ., Ylym, Ashgabat.
Babaev AG: Problems of complex development of desert and their decision in USSR (in Engl.). AS TSSR Publ., Ylym, Ashgabat.
Babaev AG and Freikin ZG: The reform of deserts in USSR. Znanie Publ., Moscow.
Veyisov S: The dynamic of relief of barkhan sands. In: Fedorovich BA and Babev AG (ed.). AS TSSR Publ., Ylym, Ashgabat.
Babaev AG (ed.): Geographic problems of Turkmenistan. Vol. III: (Collected articles). AS TSSR Publ., Ylym, Ashgabat.
Kharin NG (ed.): Geographic problems of study and development of USSR desert territories. AS TSSR Publ., Ylym, Ashgabat.
Lavrov AP (ed.): Land-water and climatic resources of desert and their rational utilization. Theses of reports of Allunion scientific conference on complex study and development of USSR desert territories. AS TSSR Publ., Ylym, Ashgabat.
Kirsta BT: Hydrological peculiarities of Central Asia Western regions (on sample of Turkmenistan). In: Shults VL (ed.). AS TSSR Publ., Ylym, Ashgabat.
Kurbanmuradov K: Types of sands of Western Low-Lying Karakum. In: Pietrov MP (ed.) AS TSSR Publ., Ylym, Ashgabat.
Lavrov AP, Larin EV and Sanin SA: Zoning of Turkmenistan takyrs for agriculture. In: Djumaev ON (ed.). AS TSSR Publ., Ylym, Ashgabat.
Nechaeva NT (ed.): Vegetation and pastures of deserts / Theses of reports of All Union scientific Conference on complex study and development of USSR desert territories. AS TSSR Publ., Ylym, Ashgabat.
Redjepbaev K and Ovsyannikov AS: Conditions of accumulation and redistribution of salt in Murgab oasis soils.
Senkevich BN: About genesis of basic forms of eolian relief of sandy deserts. In: Bayiramov S (ed.) AS TSSR Publ., Ylym, Ashgabat.
Serebryakov VP *et. al.* (ed.): Fauna and ecological-physiological peculiarities of organism in desert conditions. Theses of reports of All Union scientific conference on complex study and development of USSR desert territories. AS TSSR Publ., Ylym, Ashgabat.

1977

Aranbaev MP and Kasyev B: Microelements in soil, agricultural plants and underground waters of Murgab oasis. Vol. 1. In: Kovda VA (ed.). AS TSSR Publ., Ylym, Ashgabat.
Babaev AG (ed.): The natural environment and issues in the management of Karakum Desert. Collected works of Desert Institute. AS TSSR Publ., Ylym, Ashgabat.
Babaev AG and Freikin ZG: Deserts of the USSR, yesterday, today, tomorrow. Mysl Publ., Moscow.

Batyrov AB: The remaking land (in Turkmen language). In: Gurbanmuradov K (ed.). Turkmenistan Publ., Ashgabat.

Aranbaev MP (ed.): Geochemistry of elements and compounds in soil. AS TSSR Publ., Ylym, Ashgabat.

Kolodin MV: Desalinization of water by freezing. Klyachko VA and Meltser LZ (ed.). AS TSSR Publ., Ylym, Ashgabat.

Nechaeva NT, Shamsutdinov Z and Mukhammedov G: Improvement of Central Asia desert pastures (in English). VINITI, Moscow.

Nechaeva NT: Recommendation on improvement of pastures utilization. AS TSSR Publ., Ylym, Ashgabat.

Nikolaev VN: Organization and rational utilization of USSR desert pastures (in English). VINITI, Moscow.

Nikolaev VN, Amangeldyev AA and Smetankina VA: Desert pastures. Their fodder assessment and bonitering. In: Nechaeva NT (ed.). Nauka Publ. Moscow.

Nurberdyev M: Climate and Karakum pasture grasses. Turkmen NIINTI. Ashgabat.

Ovezliev AO, Lavrov AP: Takyrs and sands of Turkmenistan and their plant-growing development. (in English). VINITI, Moscow.

Pietrov MP: Problems of development and protection of desert nature and combat desertification (survey of literature for 1965-75). VINITI, Moscow.

Sabirova OR: Soil fauna under the sand-supporting vegetation of eastern Karakum. In: Gilyarov MS (ed.). AS TSSR Publ., Ylym, Ashgabat.

Sanin SA: Chemical and agrochemical peculiarities of Turkmenistan desert soil. In: Lavrov AP (ed.). AS TSSR Publ., Ylym, Ashgabat.

Freikin ZG (ed.): Agricultural production-territorial complex and their development. AS TSSR Publ., Ylym, Ashgabat.

Ovezliev AO (ed.): Theses of reports of the 10th scientific conference of young scientists of the Desert Institute of AS TSSR. Ashgabat.

Babaev AG: Theses of reports of scientific conference of Turkmen geographical society. Ashgabat.

Kharin NG: Remote sensing and monitoring of desertification processes in arid zones (in English). VINITI, Moscow.

Kharin NG and Pietrov MP: Glossary of terms on nature conditions and desert development. VINITI, Moscow,

Cherednichenko VP and Nuryev BN: Protection of sands from wind erosion. Turkmen NIINTI, Ashgabat.

1978

Aranbaev MP: Geochemistry of organic substance of ancient oasis soil. Kovda VA (ed.). AS TSSR Publ., Ylym, Ashgabat.

Ataev A: Efficiency of utilization of contemporary surface water flow of desert. In: Freikin ZG (ed.). AS TSSR Publ., Ylym, Ashgabat.

Ataev A, Kerimberdyev S and Nepesov B: Creation of agricultural production-territorial complexes in Central Karakum. In: Freikin ZG (ed.). AS TSSR Publ., Ylym, Ashgabat.

Nechaeva NT and Kaplin VG (ed.): Biogeocenological investigations in Eastern Karakum. Collected works of Desert Institute. AS TSSR Publ., Ylym, Ashgabat.

Gunin PD and Dedkov VP: Ecological regimes of Eastern Karakum biogeocoenosises. In: Nechaeva NT (ed.). Nauka, Moscow.
Darymova GN: Physical and chemical characteristic of Turkmenistan natural water. In: Kolodin MV (ed.). AS TSSR Publ., Ylym, Ashgabat.
Kaplin VG: Complex of soil of sandy invertebrates south subzone. In: Nechaeva NT AS TSSR Publ., Ylym, Ashgabat.
Kolodin MV: Water and life. Babaev AG (ed.). Znanie Publ. Moscow.
Lavrov AP: Soils of North-Western Turkmenistan. In: Lobova EV (ed.) AS TSSR Publ., Ylym, Ashgabat.
Saparov B (ed.): Melioration of irrigating lands in Turkmenistan. AS TSSR Publ., Ylym, Ashgabat.
Nechaeva NT, Shamsutdinov ZS and Mukhammedov GM: Improvement of Central Asia desert pastures. In: Babaev AG (ed.). AS TSSR Publ., Ylym, Ashgabat.
Nurberdyev K: Sands of piedmont plane of Central Kopetdag and their development. In: Pietrov MP (ed.). AS TSSR Publ., Ylym, Ashgabat.
Nurberdyev M: Agrometeorological conditions and productivity of Karakum pastures. Nechaeva NT (ed.). AS TSSR Publ., Ylym, Ashgabat.
Babaev AG (ed.): Repetek sandy-desert reserve. (in Russian and English). AS TSSR Publ., Ylym, Ashgabat.
Serkhenov E: Interpretation of aerophotographs of Karakum sandy massives. In: Kharin NG (ed.). AS TSSR Publ., Ylym, Ashgabat.
Rodin LE (ed.): Improvement and rational utilization of Asia zone vegetation. Nauka, Leningrad.

1979

Arnagel'dyev A: Sands of central part of Karakum, their mobility and means of combat it. In: Fedorovich BA (ed.). AS TSSR Publ., Ylym, Ashgabat.
Ataev A and Kerimberdiev S: Economical efficiency of desert pastures meliorations. Turkmen NIINTI, Ashgabat.
Mukhammedov G: Improvement of Central Karakum pastures. In: Nechaeva NT (ed.). AS TSSR Publ., Ylym, Ashgabat.
Nepesov B: Increasing of efficiency of sheep-breeding in Central Karakum. AS TSSR Publ., Ylym, Ashgabat.
Nechaeva NT, Antonova K.G. *et. al.*: Productivity of Central Karakum vegetation in connection with different regime of utilization. Nauka, Moscow.
Ovezliev AO, Dobrin LG *et. al.*: Phytomelioration of Turkmenistan deserts. In: Babaev AG (ed.). AS TSSR Publ., Ylym, Ashgabat.
Ovezliev AO (ed.): Experience of study and development of Western Turkmenistan deserts. AS TSSR Publ., Ylym, Ashgabat.
Popov KP: Pistachio in Central Asia. In: Kerbabaev B (ed.) AS TSSR Publ., Ylym, Ashgabat.
Gorelov KS *et. al.* (ed.): Problems of utilization and protection of Turkmenistan natural resources. Collected works of Desert Institute. AS TSSR Publ., Ylym, Ashgabat.
Larin EV (ed.): Results of investigations on soil science and chemicalization. Collected works of Desert Institute. AS TSSR Publ., Ylym, Ashgabat.

Prihod'ko SY, Mukhammedov GM *et. al.*: Recommendations on improvement of Kopetdag piedmont pastures. Collected works of Desert Institute. AS TSSR Publ., Ylym, Ashgabat.

1980

Aranbaev MP, Gaipova AG: Mineralogical and chemical composition of ancient-oasis soils of desert zone. In: Kovda VA (ed.). AS TSSR Publ., Ylym, Ashgabat.

Babaev AG: Desert as it is. Molodayagvardiya Publ, Moscow.

Babaev AG and Freikin AG: Alive desert. Prosveshchenie Publ, Moscow.

Batyrov A and Freikin AG: Ashgabat – 100 years. In: Babaev AG (ed.). Turkmenistan Publ, Ashgabat.

Jumashov AP: Sands of South-Western Turkmenistan and their development. In: Babaev AG (ed.). AS TSSR Publ., Ylym, Ashgabat.

Kaplin VG: Fauna of Karakum: ground vertebrates. Turkmenistan Publ, Ashgabat.

Kirsta BT: Rivers of deserts. In: Vendrov SL (ed.). AS TSSR Publ., Ylym, Ashgabat.

Kopekov CB: Natural conditions and landscape zoning of Karakum canal area. In: Babaev AG (ed.). AS TSSR Publ., Ylym, Ashgabat.

Mukhammedov G: Experience of improvement of Karakum pastures. Analytical survey. NIINTI Turkmenistan, Ashgabat.

Pakshina SM: Movement of salts in soil. Nauka, Moscow.

Svintsov IP: Recommendations on growing of planting materials of sand supporting vegetation in watering sandy nursery. AS TSSR Publ., Ylym, Ashgabat.

Mukhammedov G: Recommendations on Karakum pasture improvement. Nechaeva NT (ed.). AS TSSR Publ., Ylym, Ashgabat.

Ovezliev AO (ed.): Theses of reports of 12th scientific conference of the Desert Institute young scientists honoring 110 birthday of V.I. Lenin. AS TSSR Publ., Ylym, Ashgabat.

Fatyukhin AP: Distance methods and protection of desert nature. In: Taran IN (ed.). Nauka, Moscow.

Kharin NG: Agrophysical properties of soils and problems of their agricultural utilization. In: Nechaeva NT (ed.). AS TSSR Publ., Ylym, Ashgabat.

Nechaeva NT and Nikolaev VN (ed.): Chemical composition, nutritiousness and biological completeness of Karakum pasture fodders. AS TSSR Publ., Ylym, Ashgabat.

Khudaiyarov M: Sands of Tedjen delta and their development. In: Babaev AG (ed.). AS TSSR Publ., Ylym, Ashgabat.

1981

Babaev AG (ed.): Contemporary problems of the economic development and transformation of deserts. Collected works of the Desert Institute. AS TSSR Publ., Ylym, Ashgabat.

Ataev A: Efficiency of protection of transport and industrial objects from sandy drifts. AS TSSR Publ., Ylym, Ashgabat.

Kaplin VG: Complexes of arthropoda inhabiting in tissues of sandy desert plants. In: Nechaeva NT (ed.). AS TSSR Publ., Ylym, Ashgabat.

Karryev BB: Soils of ancient Atrek delta and means of their development. In: Charyev MK (ed.). AS TSSR Publ., Ylym, Ashgabat.

Kolodin MV: Water and deserts. In: Babaev AG (ed.). Mysl' Publ, Moscow.

Ovezliev OA and Muradov BK: Methods of plant-growing development of sands in TSSR. Turkmen NIINTI, Ashgabat.

Orlovskiyi NS: Weather and fine-fibre cotton of Turkmenistan. AS TSSR Publ., Ylym, Ashgabat.

Svintsov IP: Creation of seed-fund of Haloxylon aphyllum in Turkmenistan. Turkmen NIINTI, Ashgabat.

Charyev MK: Development of virgin lands at Lenin Karakum canal zone. Djumaev OM (ed.). AS TSSR Publ., Ylym, Ashgabat.

1982

Babaev AG (ed.) Stabilization of shifting sands of USSR deserts. AS TSSR Publ., Ylym, Ashgabat.

Kolodin MV and Sungurova MA: State and perspectives of irrigation and water-supply of deserts. Scientific survey. AS TSSR Publ., Ylym, Ashgabat.

Rabochev IC *et. al.*: Recommendations on utilization of mineralizing waters for irrigation of fodder crops in TSSR. AS TSSR Publ., Ylym, Ashgabat.

1983

Ataev A: Effectiveness of the stabilization and afforestation of sands. In: Bayiramov S (ed.). AS TSSR Publ., Ylym, Ashgabat.

Aripov EA, Nuryev BN and Arazmuradov MA: Chemical melioration of shifting sands. In: Akhmedov KS and Ivanov AP (ed.). AS TSSR Publ., Ylym, Ashgabat.

Babaev AG and Klyushkin EA: Nature reserve in Turkmenistan. AS TSSR Publ., Ylym, Ashgabat.

Ishankuliev M: Flora of Repetek biospheric reserve. In: Kalenov GS (ed.). AS TSSR Publ., Ylym, Ashgabat.

Lalymenko NK, Lalymenko II and Berdyev S: Development of takyric soils with utilization of local surface water flow. In: Babaev AG (ed.). AS TSSR Publ., Ylym, Ashgabat.

Kirsta BT: Methodical recommendations on calculation of evaporation from open surface and unirrigation soils in Turkmenistan conditions. AS TSSR Publ., Ylym, Ashgabat.

Kirsta BT: Methodical instructions on assessment of irrigation influence on water balance and some elements of hydrometeorological regime. In: Orlovskiyi NS (ed.). AS TSSR Publ., Ylym, Ashgabat.

Kharin NG, Nechaeva NT, Nikolaev VN *et. al.*: Methodical principles of study and mapping of desertification processes. In: Babaev AG (ed.). Ashgabat.

1984

Batyrov A, Muradov C et. al.: Social-economical problems of development of Tashauz oasis productive forces. In: Ataev A (ed.). Ashgabat.

Zonn IS and Orlovskiyi NS: Desertification: Combat Strategy. In: Babaev AG (ed.). AS TSSR Publ., Ylym, Ashgabat.
Babaev AG (ed.): The Desert Institute. Nauka, Moscow.
Rabochev IS (ed.): Utilization of mineralized waters in agriculture. AS TSSR Publ., Ylym, Ashgabat.
Kharin NG *et. al.*: Methodical instructions on cameral composition of pasture maps in desert zone with application of space photographs. Ashgabat.
Ovezliev AO and Agaev AA: Introduction of fodder crops in Karakum. Babaev AG (ed.). AS TSSR Publ., Ylym, Ashgabat.
Nechaeva NT and Nikolaev VN (ed.): Nutritiousness and biological completeness of North-Western Turkmenistan pasture fodders. AS TSSR Publ., Ylym, Ashgabat.
Babaev AG (ed.): Resources and natural conditions of the deserts of the USSR and their rational use. AS TSSR Publ., Ylym, Ashgabat.
Svintsov IP *et. al.*: Recommendations on stabilization and afforestation of shifting sands using protections from polyethylene films. Ashgabat.
Babaev AG (ed.): Stabilization of fodder supply of Turkmenistan deserts. Monitoring of pasture ecosystems of deserts. AS TSSR Publ., Ylym, Ashgabat.

1985

Arnagel'dyev A and Kostyukovskyi VI: Karakum desert : nature and man. In: Babaev AG (ed.). Nauka, Moscow.
Zonn IS, Orlovskiyi NS and Svintsov IP: Stabilization of shifting sands in world deserts. Turkmen NIINTI, Ashgabat
Kirsta BT and Kuznetsov NT: Water in desert. In: Babaev AG (ed.). AS TSSR Publ., Ylym, Ashgabat.
Kolodin MV: Economy of water desalinization. In: Manakov VS (ed.). Nauka, Moscow.
Krivokhatskyi VA: Insects of Repetek: cadastre species. In: Kaplin VG (ed.) AS TSSR Publ., Ylym, Ashgabat.
Lavrov AP and Orlovskyi NS: Soil-climatic zoning of Turkmenistan (outside of Karakum canal zone). In: Lobova EV (ed.). AS TSSR Publ., Ylym, Ashgabat.
Charyev MK (ed.): Development of clay soils of Turkmenistan. Methodical recommendations. AS TSSR Publ., Ylym, Ashgabat.
Babaev AG (ed.): Essays of Turkmenistan nature and economy. Collected works of Desert Institute. AS TSSR Publ., Ylym, Ashgabat.
Nechaeva NT (ed.): Improvement of Soviet Central Asia desert pastures. (in English). AS TSSR Publ., Ylym, Ashgabat.
Kharin NG *et. al.*: Methodical principles of study and mapping of desertification processes. In: Babaev AG (ed.) (in English). Ashgabat.
Cherednichenko VP and Darymov VY: Geomorphological principles of industrial development of Turkmenistan sandy deserts. In: Babaev AG (ed.). AS TSSR Publ., Ylym, Ashgabat.
Shirmamedov M: Western Turkmenistan shifting sands and their development. In: Ovezliev AO. AS TSSR Publ., Ylym, Ashgabat.

1986

Babaev AG (ed.): The use of aerial photographs and satellite imagery for analyzing the natural condition of deserts. AS TSSR Publ., Ylym, Ashgabat.

Babaev AG, Zonn IS, Drozdov NN and Frieikin ZS: Deserts. In: Murzaev EM (ed.). Mysl' Publ., Moscow.

Kirsta YB: Modelling of desert ecosystems. In: Nechaeva NT (ed.). AS TSSR Publ., Ylym, Ashgabat.

Manakov VS (ed.): Scientific-technical progress in desert. Theses of reports of 5th All Union scientific conference. AS TSSR Publ., Ylym, Ashgabat.

Nechaeva NT and Nikolaev VN (ed.): Natural resources of deserts and their development. Theses of reports of 5th scientific conference. AS TSSR Publ., Ylym, Ashgabat.

Prikhod'ko SY and Prikhod'ko ND: Phytomelioration of Turkmenistan desert pastures. Turkmen NIINTI, Ashgabat.

Nechaeva NT and Nikolaev VN (ed.): Chemical composition, nutritiousness and biological completeness of pasture fodder of Turkmenistan's piedmont regions. AS TSSR Publ., Ylym, Ashgabat.

Babaev AG and Kharin NG: Ecological problems of desert development and protection of nature. Theses of reports of 5th All union scientific conference (Ashgabat, Okt., 1986). AS TSSR Publ., Ylym, Ashgabat.

1987

Kirsta BT: Peculiarities of calculation of hydrometeorological characteristics of Turkmenistan. In: Kuznetsov NT (ed.). AS TSSR Publ., Ylym, Ashgabat.

Redjepbaev K and Esenov P: Changing of soil-meliorative conditions of Khauzkhan massif in connection with irrigation development. In: Rabochev IS (ed.) AS TSSR Publ., Ylym, Ashgabat.

Babaev AG (ed.): Ecological and geographical aspects of desert peculiarities. Theses of reports of 16th scientific-practical conference of young scientists. AS TSSR Publ., Ylym, Ashgabat.

1988

Arnagel'dyev A, Kostyukovskyi VI: Ecosystems of Karakum. In: Babaev AG (ed.). AS TSSR Publ., Ylym, Ashgabat.

Kirsta BT and Orlovskyi NS (ed.): Peculiarities of water-warming regime and agroprognosis in Turkmenistan. Collected articles. AS TSSR Publ., Ylym, Ashgabat.

Svintsov IP: Forest melioration of Turkmenistan sandy desert. Ovezliev AO (ed.). AS TSSR Publ., Ylym, Ashgabat.

Kharin NG: Methodical manual for translating of literature on distance methods and study of nature resources. In: Babaev AG (ed.). AS TSSR Publ., Ylym, Ashgabat.

Kharin NG: Through the Libya. In: Babaev AG. AS TSSR Publ., Ylym, Ashgabat.

Charyev MK and Garagulov MC: Development of Turkmenistan virgin lands. Practical recommendations. AS TSSR Publ., Ylym, Ashgabat.

1989

Aranbaeva EV: Organic substance of anthropogenic – oasis soils of desert zone. AS TSSR Publ., Ylym, Ashgabat.

Aripov EA and Nuryev BN: Physical and chemical mechanics of shifting sands. Fan Publ., Tashkent.

Batyrov AB, Prikhod'ko LN and Petrosyan LM: Territorial – resource peculiarities of Turkmenistan desert development. Turkmen NIINTI, Ashgabat.

Gorelov SK, Jumashov AP and Mokrushina LS Zaunguz Karakum development of relief and problems of rational land use. Moscow.

Ivanov AP: Formation of profiles of eolian relief forms of sandy deserts. In: Orlovskyi NS (ed.). AS TSSR Publ., Ylym, Ashgabat.

Rabochev GI and Pyagai ET: Changing of sandy soils properties under the influence of irrigation by mineralized water. In: Sanin SA (ed.). AS TSSR Publ., Ylym, Ashgabat.

Kharin NG, Dmitriev MG, Orlovskyi NS *et. al.*: Arid Center of collection and processing of space information (conception, technical, information and program provision). In: Babaev AG (ed.): Methodical instructions of the organization of the center. AS TSSR Publ., Ylym, Ashgabat.

Esenov P: Meliorative assessment of Tashauz oblast irrigation and drainage waters, their rational utilization and protection. In: Rabochev GI (ed.). AS TSSR Publ., Ylym, Ashgabat.

1990

Arnagel'dyev A: Morphodynamics of the wind erosion relief in sandy deserts. In: Ovezliev AO (ed.). AS TSSR Publ., Ylym, Ashgabat.

Arnagel'dyev A and Kostyukovskyi V: Deserts. Rational exploitation and protection. In: Babaev AG (ed.). Agropromizdat, Moscow.

Babaev AG and Gorelov SK: Problems of desert geomorphology. In: Orlovskyi NS (ed.). AS TSSR Publ., Ylym, Ashgabat.

Jumashov AP: Ecological and geographical conditions and types of Central Asia deserts. In: Babaev AG (ed.). AS TSSR Publ., Ylym, Ashgabat.

Batyrov AB (ed.): Scientific principles of population distribution and economic development in arid regions. AS TSSR Publ., Ylym, Ashgabat.

Ovezliev AO, Ibragimov AS and Semergei KI: Fertilizing of maize in Karakum. In: Sanin SA (ed.). AS TSSR Publ., Ylym, Ashgabat.

Sabirova OR: Soil mezofauna of Turkmenistan deserts. In: Kaplin VG (ed.). AS TSSR Publ., Ylym, Ashgabat.

1991

Babaev AG: Geo-ecological analysis of the dynamics of desert geosystems dynamics using remote sensing. In: Kharin NG (ed.). AS TSSR Publ., Ylym, Ashgabat.

Jarkov VV, Utyaganov RZ and Jarkov DV: Sorption cleaning of collector drainage water. Turkmen NIINTI, Ashgabat.

Kirsta BT: Resources of surface waters and natural moistening of Turkmenistan territories. Kuznetsov NT (ed.). AS TSSR Publ., Ylym, Ashgabat.

Nechaeva NT and Mukhammedov G: Monitoring of natural and improving vegetation of Central Karakum. In: Nikolaev VN (ed.). AS TSSR Publ., Ylym, Ashgabat.

Svintsov IP: Combating desertification in the Sahel region of the Republic of Mali. Ovezliev AO (ed.). AS TSSR Publ., Ylym, Ashgabat.

Seyitkurbanov S: Combining helio-wind power installations. In: Bairamov RB (ed.). AS TSSR Publ., Ylym, Ashgabat.

Kharin NG: The Gobi and man: popular scientific edition. In: Babaev AG (ed.). AS TSSR Publ., Ylym, Ashgabat.

Nechaeva NT and Nikolaev VN (ed.): Chemical composition, nutritiousness and biological completeness of pasture fodders of South-Western Turkmenistan. AS TSSR Publ., Ylym, Ashgabat.

Cherednichenko VP: Engineering-geomorphological investigations of Karakum. In: Babaev AG (ed.). AS TSSR Publ., Ylym, Ashgabat.

1992

Ataev SA: Natural meliorative conditions of South-Western Turkmenistan. In: Babaev AG (eg.). AS TSSR Publ., Ylym, Ashgabat.

Babaev AG *et. al.* (ed.): Karakum desert and Thar desert. AS TSSR Publ., Ylym, Ashgabat.

Jarkov VV, Jarkov DV and Filipova AV: Domestic apparatus for drinking water cleaning: "Cheshme". Turkmen NIINTI, Ashgabat.

Kolodin MV: Water-protection measures in arid zone. In: Babaev AG (ed.). AS TSSR Publ., Ylym, Ashgabat.

Lalymenko NK and Lalymenko LA: Growing of fodder crops in arid zone by irrigation with economic-domestic sewage. In: Orlovskyi NS (ed.). AS TSSR Publ., Ylym, Ashgabat.

Nurberdyev M and Reizvikh ON: Productivity of Central Asia desert pastures. In: Nikolaev VN (ed.). AS TSSR Publ., Ylym, Ashgabat.

Svintsov IP, Yurchenko NN *et. al.*: Strong swelling polymeric preparations (SPG) and their utilization for sands melioration. Ashgabat.

Shoshov LL and Aranbaev MP: Genesis, classification and agricultural using of arid subtropics soils and their mountain setting. AS TSSR Publ., Ylym, Ashgabat.

1993

Arnagel'dyev A and Kurbanov OR: Study of native land nature. Babaev AG (ed.). (in Turkmen language). AS TSSR Publ., Ylym, Ashgabat.

Babaev AG, Kharin NG and Orlovskyi NS: Assessment and mapping of desertification processes: a methodological guide. Ashgabat.

Batyrov AB: Deserts: peculiarity of development and settling. In: Lappo GM (ed.). AS TSSR Publ., Ylym, Ashgabat.

Kaplin VG: Open living arthropoda of Karakum seed plants. In: Nechaeva NT (ed.). AS TSSR Publ., Ylym, Ashgabat.

Nechaeva NT: Half a century in Karakum. In: Babaev AG (ed.). AS TSSR Publ., Ylym, Ashgabat.
Nechaeva NT, Mukhammedov GM and Shamsutdinov ZS: Enrichment of Central Asia arid pastures. In: Babaev AG (ed.). AS TSSR Publ., Ylym, Ashgabat.
Orlovskyi NS, Fatkulin MN *et. al.*: Air temperature on a global scale and magnetic activity. Izmiran, Moscow.
Orlovskyi NS, Fatkulin MN *et. al.*: Air temperature on a global scale and solar activity. Izmiran, Moscow.
Kharin NG, Kalenov GS *et. al.*: Explanatory note to map of anthropogenic degradation of soils in Aral Sea basin. In: Babaev AG and Kharin NG. AS TSSR. Center of invest. ESCATO/UNEP. Ashgabat.
Kharin NG, Kiril'tseva AA and Gringof IG: Season natural phenomena. Gidrometeoizdat, Sankt-Peterburg.
Kharin NG, Orlovskyi NS *et. al.*: Criteria and methodology of assessment of desertification processes in the Sahel. Babaev AG (ed.). AS TSSR Publ., Ylym, Ashgabat.

1994

Amanniyazov KN: Geology and mineral wealth of Turkmenistan. AS TSSR Publ., Ylym, Ashgabat.
Amanniyazov KN and Fedin VP: Garabogazgol - second birth. AS TSSR Publ., Ylym, Ashgabat.
Babaev AG and Zonn IS: Desertification problems: international cooperation. In: Orlovskyi NS (ed.). AS TSSR Publ., Ylym, Ashgabat.
Gurbanov OR and Arnagel'dyev A: Study of land nature (in Turkmen language). In: Babaev AG (ed.). AS TSSR Publ., Ylym, Ashgabat.

1995

Babaev AG: Problems of desert development. Global, regional and national aspects. Zonn IS (ed.). AS TSSR Publ., Ylym, Ashgabat.
Jarkov VV and Saparov AG: Drinking water today and tomorrow. Garlavach, Ashgabat.

1996

Babaev AG: Problems of arid land development (in English). Moscow University Press., Moscow.
Ilamanov A, Babaev AG and Batyrov AB: Every drop - a little gold (in Turkmen language). Turkmenistan Publ., Ashgabat.

Plate 2.1. *Haloxylon persicum*

Plate 3.1. Agricultural Development of oasis sands

Plate 6.1. Satellite image got from „IRS-IA“

Plate 6.2. Satellite image taken from "LandSat-2" (600-700 nm) of Murgab territory and part of Tedjen oasis
(1) arable land
(2) Tamarix florida plants
(3) desertification spots
(4) reservoirs
(5) Karakum canal
(6) railway
(7) takyrs
(8) filtration lakes out of Karakum canal
(9) rivers
(10) sands.

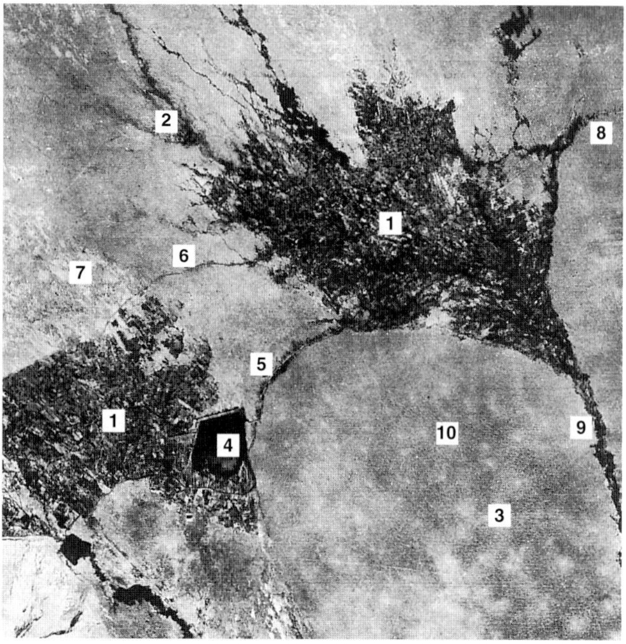

Plate 6.3. Aerial photo of sand hills (severe) and faintly anchored sands (moderate degree of desertification)

Plate 6.4. Aerial photo of cellular sands in Karakum desert illustrating a moderate desertification rate

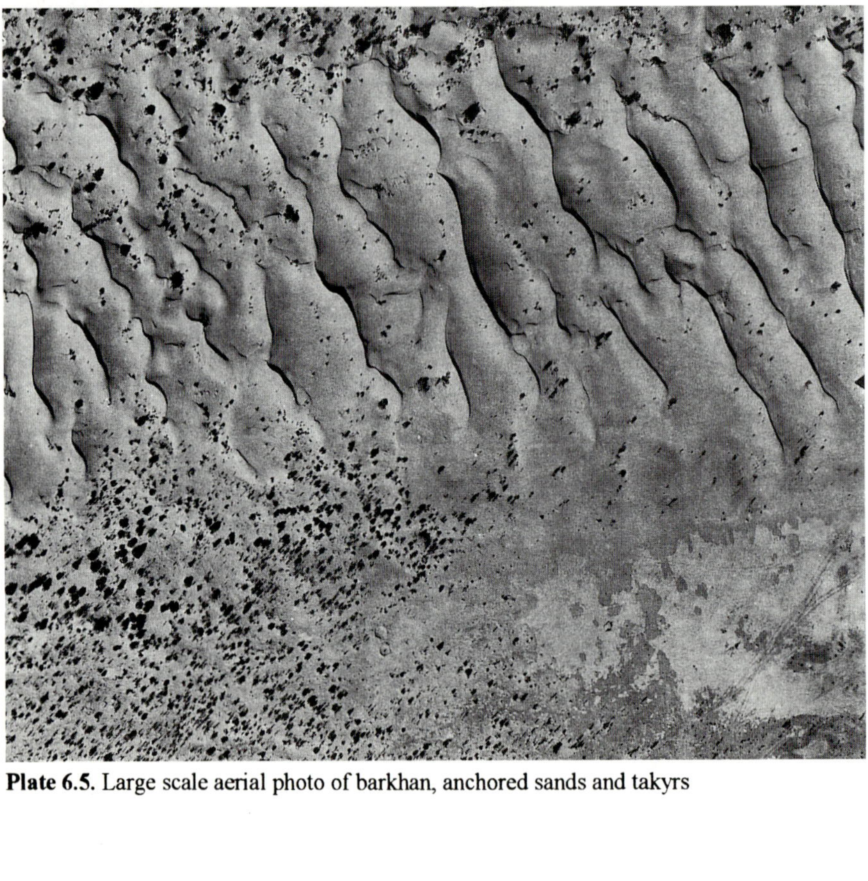

Plate 6.5. Large scale aerial photo of barkhan, anchored sands and takyrs

Plate 6.6. Aerial photo of Karakum canal, filtering lakes and cellular sands

Plate 6.7. Aerial photo of barkhan, sands formed around settlement areas in a desert due to overgrazing and deforestation (severe desertification) and takyrs places with water erosion process, caused by surface flowing

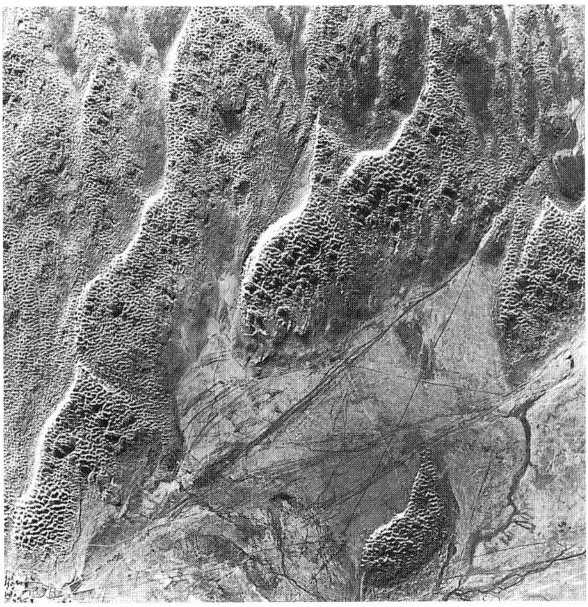

Plate 7.1. Bare sands

Plate 7.2. Overgrown sands

Plate 7.3. Loess desert

Plate 7.4. Takyr in clay desert

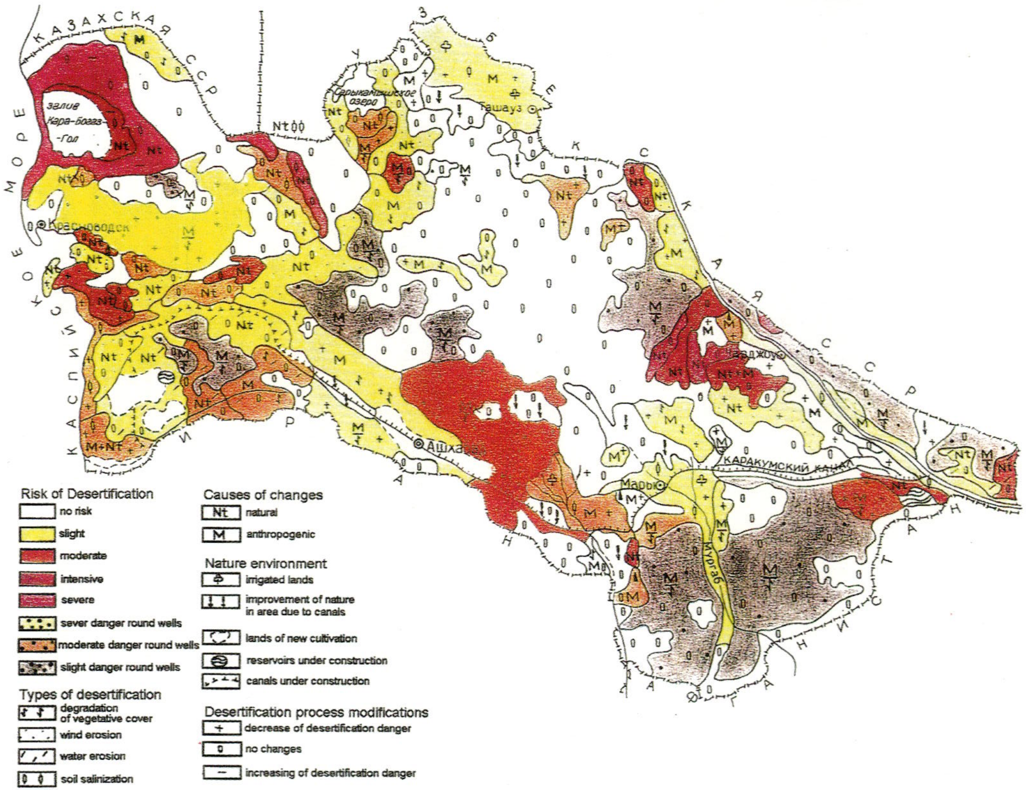

Plate 8.1. Fragment of the *"Map of desertification risk for the next 20 years"*

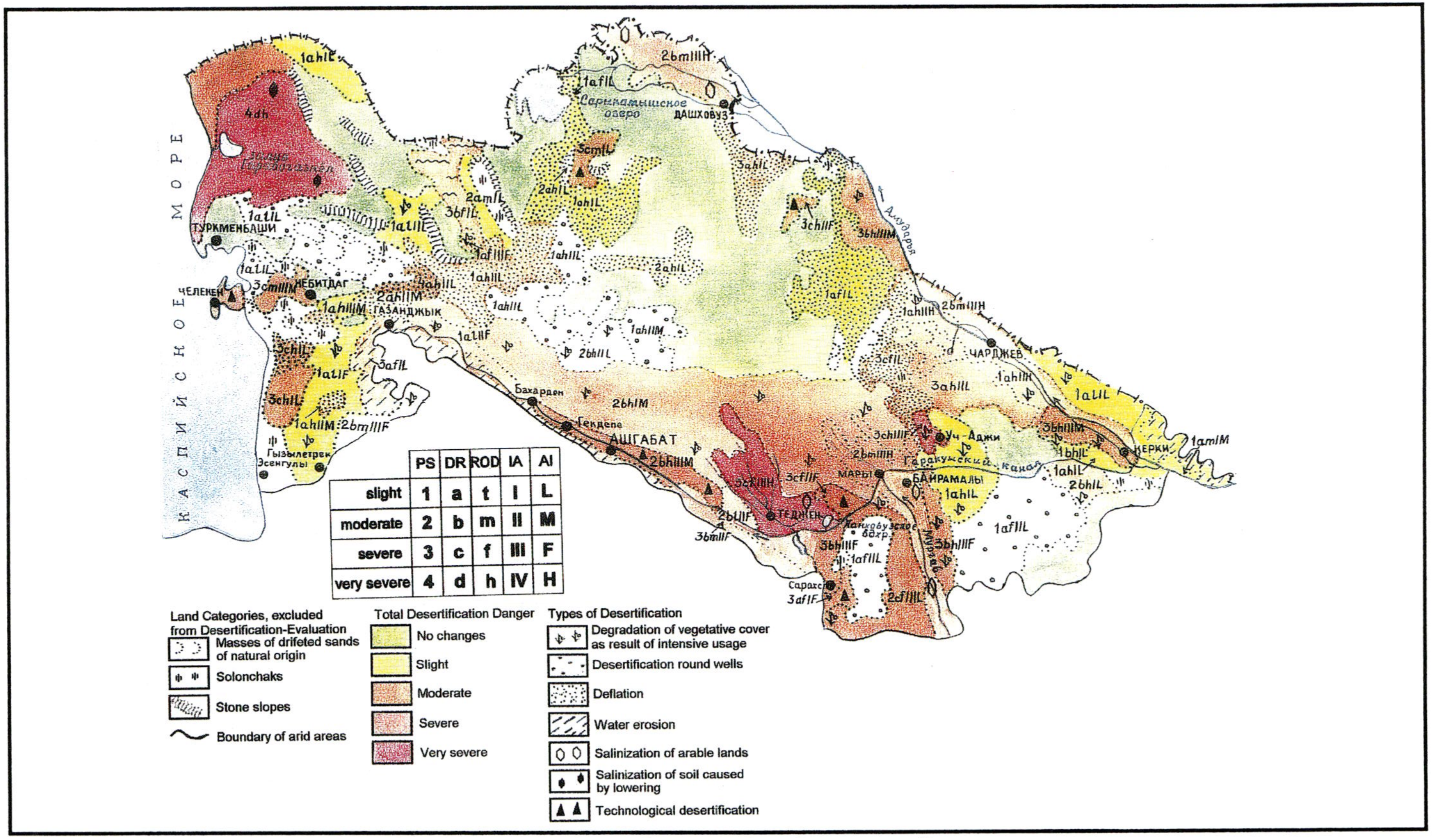

Plate 8.2. Fragment of the *"Map of anthropogenic desertification of arid territories"*

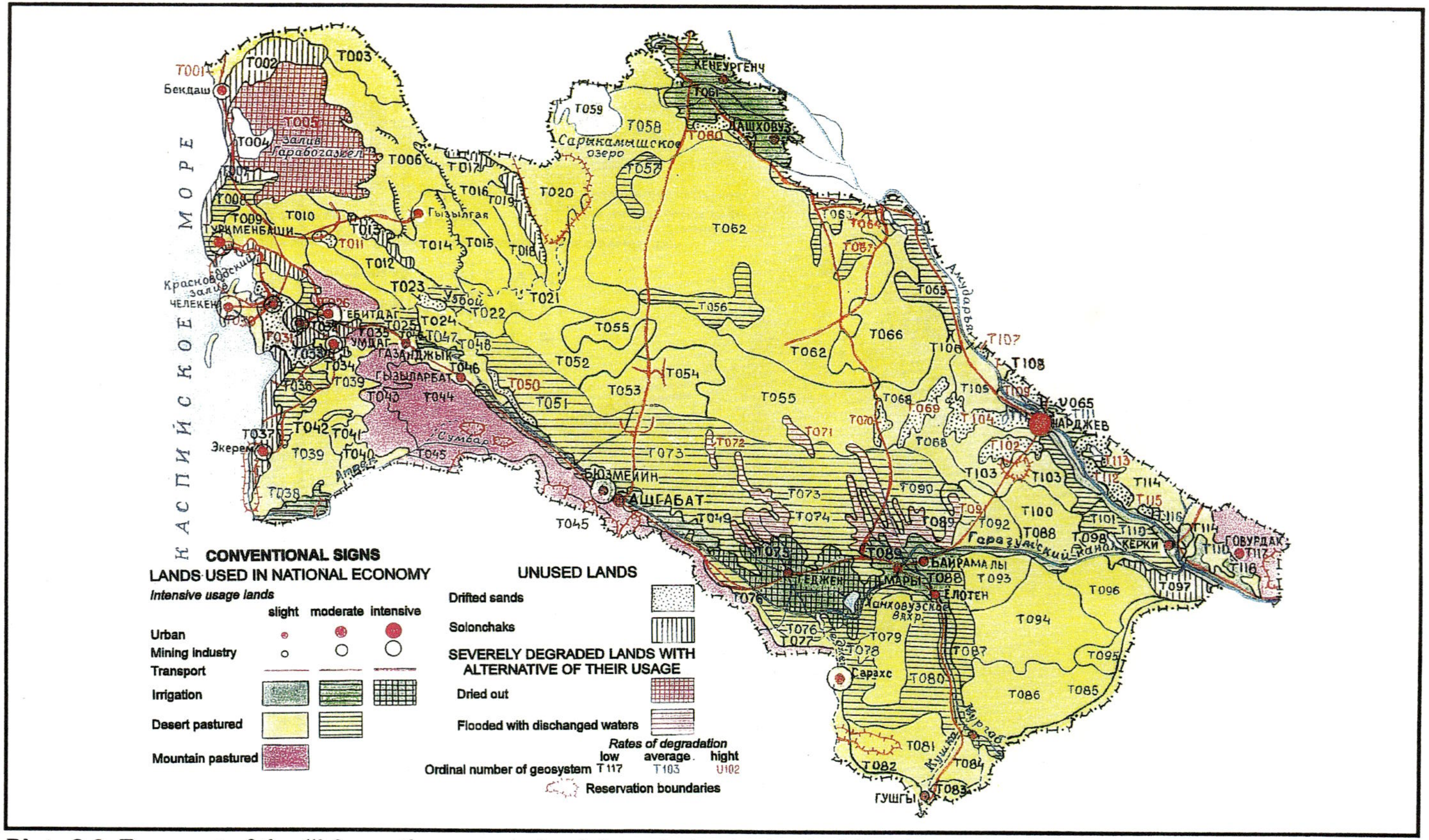

Plate 8.3. Fragment of the *"Map anthropogenic land degradation in the Aral Sea Basin*

Plate 9.1. Year-round pastures created by sowing of seeds of shrubs, semishrubs and grasses in the piedmont desert

Plate 9.2. Shrub pastures created by planting of seedlings on barkhan sands

Plate 9.3. Pastures with *Haloxylon aphyllum* created by sowing of seeds on embedded hillock sands

Plate 12.1. Apple trees on sands, watered by overhead irrigation

Plate 12.2. Maize "Dneprovskaya" on sands

Plate 13.1. Squarely arranged vegetation as road protection from moving sand

Plate 13.2. Fixation of sandy surface by means of bitumen solution and transformer's oils

Plate 14.1. A furrow of a double-breasted cross-section cut by trencher

Plate 15.1. Industrial cotton-cropping on irrigated salinized land. In wide parts of the field crops are not growing because of salinization

Plate 15.2. Salinized fallow of formerly irrigated land

Plate 22.1. Valley-like depressions with thickets of *Haloxylon persicum* in Repetek International Biosphere Reserve

Plate 22.2. *Haloxylon persicum* growing using slightly saline groundwater

Springer and the environment

At Springer we firmly believe that an international science publisher has a special obligation to the environment, and our corporate policies consistently reflect this conviction.
We also expect our business partners – paper mills, printers, packaging manufacturers, etc. – to commit themselves to using materials and production processes that do not harm the environment. The paper in this book is made from low- or no-chlorine pulp and is acid free, in conformance with international standards for paper permanency.

Springer

Druck: Druckhaus Beltz, Hemsbach
Verarbeitung: Buchbinderei Schäffer, Grünstadt